HANDBOOK OF ADVANCED ROBOTICS

Other TAB books by the author:

HANDBOOK OF
ADVANCED
ROBOTICS

BY EDWARD L. SAFFORD, JR.

TAB TAB BOOKS Inc.

BLUE RIDGE SUMMIT, PA. 17214

FIRST EDITION

SECOND PRINTING

Copyright © 1982 by TAB BOOKS Inc.

Printed in the United States of America

Library of Congress Cataloging in Publication Data

Safford, Edward L.
    Handbook of advanced robotics.
    Includes index.
    1. Automata.   I. Title.
TJ211.S24   629.8′92   82-5697
ISBN 0-8306-2521-6   AACR2
ISBN 0-8306-1421-4 (pbk.)

Contents

To my Parents.

Introduction

Robotics is more than just a word, it is a reality. It means, essentially, the doing of something with computer-controlled machines. These machines are different. They are different as people are different and are given to different likes and dislikes, tasks and interests. There are some who say that robotics is an art, the art of getting the impossible from machines. There are others who say that robotics is a science, the science of doing with uncanny quickness and precision, tasks which we as humans can do, but don't like to do, or find it very boring to do. Then, there is the impossible. In this category we find chores where breathing is impossible, or where we would burn up readily.

Recall a machine that cleans the hulls of large ships: a scrubber. It is equipped with fast rotating brushes of special alloys and a gripping force to cause the scrubber to adhere to the side of the ship under water. That little scrubber does all its cleaning automatically, under programmed instructions from its computerized brain.

There is a story about a man on an automobile assembly line. He was a good worker, intelligent, and aggressive. He became bored with the task of continually inserting bolts into pre-drilled holes, hour after hour. One day he decided not to insert the bolts as he had been inserting them in the past—just to give his task some interest and variation. He cross-threaded them, used short bolts and sometimes didn't use any bolts at all. Your imagination can fill out the further details of this story and what interesting and imaginative things you can come up with as you think about the various people who bought those cars!

Robots can scrub or put bolts in holes forever—assuming that the power is kept on and the mechanics don't fail. Robots do not get tired or bored or even interested in what is going on. They just do what their little computerized brains tell them to do. Of course, some robots have really big brains but require knowledgeable, scientific human types to keep them operating smoothly.

You will, no doubt, immediately think of other applications for these machines. Although we hate to admit it, we can come up with one more reason for having a machine do tasks for us. *They can do it better!* Yes, it's true, they can do it better, for a longer period of time, and more consistently than any human can ever do. You see they have one advantage—they do not get tired!

If we ask the fundamental question, will machines and computers take over the world? We will find that we've a disagreement on our hands. Some look at the exploding field of robotics and say there is no doubt that someday machines will control everything. These persons worry about jobs and activities and—yes, it is true—a loss of control over the machines by the humans who make them. They envision that someday machines will be making machines which control machines, etc. *ad infinitum.*

One thing is sure, no one has ever seen a machine making love, so there may be hope. Indeed, there *is* hope for all of us for an ever increasingly pleasurable future due to the fact that machines are around and doing things for us.

I extend my gratitude to the many companies and individuals who provided me with information, discussions, questions, and answers for this book. Their illustrations are most informative. Now, for a trip through the fascinating world of *advanced robotics,* turn this page and keep going.

Acknowledgments

Thanks to the following companies for their cooperation in making this book possible.

American Robot, P.O. Box 10767, Winston-Salem, NC, 27108

PRAB Conveyors, Inc., 5944 East Kilgore Road, Kalamazoo, MI, 49003

Advanced Robotics, Newark Industrial Park, Hebron, OH

Texas Instruments, P.O. Box 1443, Houston, TX 77001

Quasar Industries, 59 Meadow Road, Rutherford, NJ 07070

Heuristics, Inc., 900 San Antonio Road, Los Altos, CA, 94022

Solfan, 665 Clyde Avenue, Mountain View, CA, 94043

Jerry Rebman Electronics, 5439 Doliver, Houston, TX, 77027

Robotics Today, The Strand Building, Peterborough, NH 03458

Cincinnati Milacron, 4701 Marburg Avenue, Cinnati, OH 45209

ASEA Electronics Division, Vasteras, Sweden

General Electric, Optoelectronic Systems, Electronics Park, Syracuse, NY 13201

Spatial Data Systems, 508 Fairview Drive, Goleta, CA, 93017

Polaroid, Ultrosonic Ranging Marketing, 784 Memorial Drive, Cambridge, MA, 02139

Mountain Hardware, 300 Harvey West Boulevard, Santa Cruz, CA 95060

Ohio Scientific, 1355 South Chillocothe Road, Aurora, OH 44202

Radio Shack, a division of Tandy Corporation, Fort Worth, TX,

Microbot, 1259 El Camino Real, Suite 200, Menlo Park, CA 94025

Manca, Inc., Leitz Building, Rockleigh, NJ 07647

Camac Kenetic Systems, 11 Maryknoll Drive, Lockport, IL 60441

Kilobaud Microcomputing, 73 Pine Street, Peterborough, NY, 03458

ACE Radio Control, Higginsville, MO, 64037

Telesensory, Inc., P.O. Box 10099, Palo Alto, CA 94304

Planet Corporation, 27888 Orchard Lake Road, Farmington Hills, MI, 48018

Robotics Age, P.O. Box 725, La Canada, CA 91011

Apple Computer, Inc., 10200 Bandley Drive, Cupertina, CA, 95014

Definitions, Examples, and Ideas

What is robotics? This is not a new word but it may be an unfamiliar one. Automation is another somewhat similar word that applies in many instances to the same idea or concept. It is only through the magic of the big screen TV and science fiction stories that the word robot has come into such prominence. It is now almost a household word. The word robot has come to be associated with the C3POs of other worlds; those mechanical, man-like units of metal and electronics that play roles in stories of the future. It is interesting that the word robot can also mean an unthinking robot being who performs some task automatically. It is not considered normal for a human because that is the way *machines* are supposed to operate.

Machines act and do not think—or at least do not think or reason as humans think and reason.

Robots do have memory banks that are usually programmed in some manner. They do speak and seem able to understand human speech. Because we have so many varieties of solid-state circuits nowadays, no one is surprised when a robot speaks, or a calculator, game, TV or automobile speaks in a distinct monotone, but remember, they don't think!

The word robot comes from the Czechoslovakian word *robotnik* meaning slave, servant, or compulsory service. The word has been defined as being descriptive of an automaton or a person who acts or works mechanically without thinking for himself. By definition, a robot pilot is a device which serves as an automatic pilot in an airplane for example. Note that the word automatic appears in the definition.

During World War II the concept of an automatic pilot was extended to guided missiles (also called robot bombs). It kept them flying upright and capable of executing steering commands sent from an automatic ground computer-controller. This points to another link in the system—the data or command transmission link.

The exploration of space has provided robots of a much more sophisticated variety. These mechanical marvels seem to be able to think and act on their own as they travel into the vastness of the cosmos. These robots have the ability to sense situations and conditions. Automatically they extend measuring or photographic instruments from protective housings and sample the ether, or planets, or asteroids. They place the information in memory banks and then send it, slowly, allowing plenty of time for accurate transmission, back to the controller-computer on earth where the data is assembled, enhanced, and displayed for human consideration and amazement. When so programmed, these robot marvels steer themselves down to a planet's surface. They land carefully, adjust themselves, and communicate their status, position, and findings, in this new environment. It has been rumored that the Russians have developed this type robot to such a state of perfection that they land on the moon, take samples, lift off, and return to earth without the assistance of human control in flight or during on-planet actions. They act like the mechanical men of Karel Copek's play *R.U.R.* even if they do not have humanoid appearance.

There can be no doubt that man is moving toward that day when some type of machine will be used in almost every human capacity. The scientists believe that people would like to have this machine look somewhat human, and so they spend considerable time in an effort to develop mechanical eyes, ears, vocal chords, arms, hands, legs, and a method of locomotion which simulates human walking. Everyone is delighted when a mechanical machine, looking very much like a Middle Ages knight-in-armour clumps around and performs some act of a semi-useful nature, or utters sounds which are human words, all in proper response to voiced human commands.

While it all seems relatively normal, response of a machine to human commands uttered vocally is not simple. It is the world of *algorithms* which is what the scientists and engineers call their magic equations. Algorithms instruct and control the machine and everything it does. Associated with every type of mechanical device which does anything, there is some type of controlling device which reads an algorithm. In order to understand and communicate directly, we should also have some understanding of algorithms which are used in both operational and developmental situations to make things come alive.

When we consider all the parts of the world of robotics, we are talking about robotics. In the Introduction I have described this as being the art or science of controlling machines such that they do something. We now consider what that *something* might include and thus consider in more detail what we might mean by saying that robotics is an art.

Art is defined as a skill acquired by experience or study. It also has been defined as the systematic use of knowledge or skill in making or doing things, and the use of skill and imagination in the production of things of beauty, or the things produced which are so considered. So it seems from these definitions that robotics means machines, computers, mathematics, algorithms, science, engineering, and some artistry. If we try to come up with a simple kind of definition for robotics we might try: The design, use, and operation of machines, which are computer controlled by algorithm, to do human-desired tasks.

This definition does not include any specification as to the shape or size or type of machine, nor does it limit itself to a definition of a physical output—although that kind of output is usually associated with the word robot. Notice also that we have indicated the machine must be controlled by an algorithm and implied in this definition is some type of computer. We try not to limit the machine to either fixed or mobile construction, and we do not limit the machine to an autonomous operating condition. You may take the extreme position that the human brain is also a computer! This means that waldos are also robots.

The incorporation of a human's brain somewhere within the flow pattern of operational commands for a mechanized unit has to be considered at some stage, even though it may be by-passed at later stages and in subsequent operations of the electromechanically directed unit. If you ask what this means, I would respond, "We are referring to the *teaching stage* in development of a system which will then become autonomous."

"What about self learning?" you might ask. That comes under the heading of *adaptive control*, which simply means that the machine can adapt itself to changing conditions of input that determine its operational output. In this case the machine is equipped with sensors that can recognize changes in the position of objects that it is to pick up and convey to another place. As it determines these changes, its sensor input can cause changes or modifications to its control-computer program so that the machine's mechanical movements can be modified to adjust its pick up arm or hand or gripper and continue to do its task without fault. Adaptive control capability is still controlled by the original algorithm, though.

The ability of a machine to do things of a constructive nature depends upon the accuracy, sensitivity, number, and type of it's sensors. Consider for a moment the human ability to pick up a pencil. The sensors involved would be the eyes, which determine at every instant the relative position of the hand, arm, and fingers to the pencil, and the skin sensors, which feel the pencil when it is touched. The sense of feel allows the brain to determine the tightness of grip and position of the grip necessary to pick up the pencil. The eyes help the brain determine where to move the pencil. If the underlying command of operation was "Pick up the pencil and write 50 on a paper" the eyes would also have determined the position of the paper, the skin sensors would have determined that the paper and pencil met under the proper circumstances, and then the algorithm causing the fingers to make the movements to write 50 would have been initiated.

What the human machine is getting throughout this operation is *feedback* from all the various sensors necessary for this kind of operation. Sensors provide input signals of what the hand is doing. What the hand is doing is the machine's output. When a machine inputs signals from its own output this is called feedback, and this is vital to the operation of any machine considered under the heading of robotics. Feedback is a concept long known in engineering circles and in the world of servomechanisms. What is new in the world of robotics is the type, size, and accuracy of the sensors, and the ability to compute very complex equations very quickly to determine exact operational signals for the robot.

The human body is one big mass of sensors. Almost every point on the skin's surface is sensitive to heat and pressure in ranges from the smallest puff of wind to the largest rib-smashing blow, and from a bitter winter morning to a humid summer scorcher. Add to this the hearing sensors, the visual sensors, and the taste and smell sensors and you have a pretty complex and sensitive package. To enrich your thinking along this line, consider that pencil picked up a few paragraphs ago. How does the brain know that it is a pencil when the eye sensors see it? If you answer that the brain has a stored image of a long slender multisided object with a sharp point, then you are getting close to the kind of thinking used to define algorithms in robotics. Think of how many images our brains must have in storage on a relatively permanent retention basis, and how it discards old images and incorporates new images. Then consider this thought provoking definition: Robotics is the science of making machines man-like in action and operation.

To discuss a robot is one thing, to examine it is another. Let's take a look at one kind of robot. The Grivet Series 5 industrial robot in Fig. 1-1 might be a surprise because it consists of a single automated arm that performs under the direction of a control unit, the TARC (Fig. 1-2).

The Grivet responds to the teaching instructions of the person holding the teaching box in Fig. 1-1. We are assured that teaching this kind of robot to do important industrial tasks is easy. In fact, the American Robot Corp. says that this unit is designed to be extremely simple to use and to maintain. What about maintenance? Robotic devices are electromechanical machines and thus their ability to function and to keep functioning for long unattended periods of time depends on how the designer planned for maintenance—and possibly troubleshooting, or diagnostics.

I am reminded of a situation concerning the development of a new guided missile robot system. Many manufacturers and scientists were present and deeply involved in the planning for the system. When it came time to consider maintenance and repair, everyone agreed that the system would "almost never" break down or need such attention. Just in case it should, a testing unit was to be devised which would automatically run tests and isolate troubles so quick and effective repairs could be made. You guessed it! *The complexity of the testing unit was of such magnitude that it became advisable to consider a testing unit to test this testing unit!*

We want our robot systems to be as free from troubles and maintenance as possible, but what testing and maintenance they have should be easy to perform, consistent in application, and totally effective in operation.

Look back at Fig. 1-1. This Grivet robot has been designed for industrial applications where a job handling objects weighing less than three pounds is the primary concern. It can be taught what to do for a particular task with just a few minutes of instruction, using the hand held teaching unit. A close-up of this unit is shown in Fig. 1-3. Simplicity may be a key to easy maintenance.

TEACHING A ROBOT

Robot Master John Galaher of the American Robot Corp. informs us that this box controls extension and contraction of

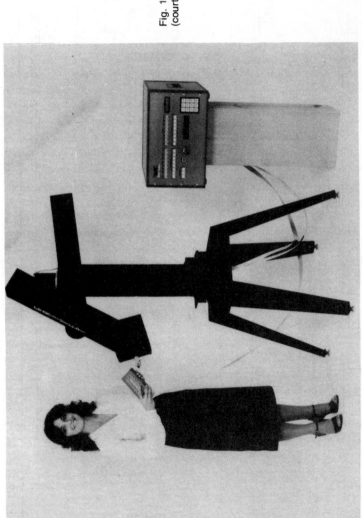

Fig. 1-1. The Grivet Series 5 Robot (courtesy American Robot Corp.).

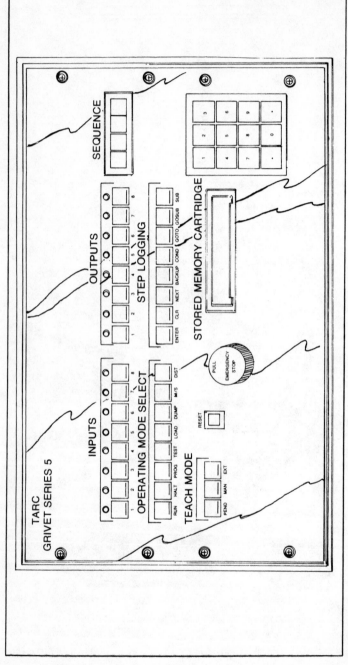

Fig. 1-2. The TARC control unit for the Grivet Robot (courtesy American Robot Corp.).

8

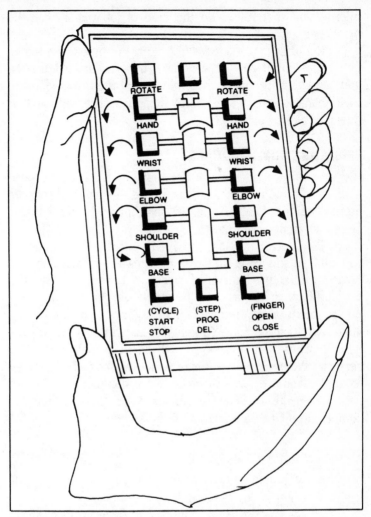

Fig. 1-3. The teaching unit (pendant) for the Grivet robot (courtesy American Robot Corp.).

the arm and rotation of the hand gripper, wrist, elbow, and shoulder. One sets this kind of robot into its permanent location, programs it to accomplish the task desired, and then removes the teaching unit. The arm, under direction of TARC, will now go to work doing what you have told it to do. In this case it may replace a person on a batch separation line, or change objects from one conveyor belt to another. It is man-like in that it can do the kind of job that a man did previously.

"Robotics is the science of making machines man-like in action and operation." This seems to be an accurate definition and you will note that it does *not* say that the machine has to look man-like (or woman-like) in appearance, although it doesn't rule it out either!

EMULATING HUMAN ACTIONS

In many discussions of the capabilities of a robot arm such as the one in Fig. 1-1, it is stated the arm will *emulate* human actions. Perhaps you are familiar with this word. It seems an important one.

"To emulate" means to strive to excel, or to exceed, or to equal. In this case the arm emulates the task which was formerly done by a human. Of course we have to consider the possibility that the task will be a new one, or one in a new environment or location which has never been performed by anyone, human or not. We imagine that production will require a whole list of tasks, resulting in a final product or service. We look at the performance of these tasks as an example, and we have to make some comparisons, so we naturally make the comparison between the machine accomplishment and a human accomplishment.

GRIVET-ROBOT ARM DYNAMICS

Let's examine that Grivet robot arm in more detail using Fig. 1-4 to guide our thinking.

In A we see the static condition of the arm. It is tightly positioned against its base, and it does nothing. In B it has been sent signals to extend, and from the illustration we can see through what angles rotary motion can be accomplished, and also via these same angles a translatory motion is made for the gripper. This is easy to understand if you imagine that section 1 of the arm moves or rotates downward, while section 2 moves or rotates back, and section 3, the wrist, rotates downward to remain on a level line from the starting position. Notice also that the arm itself can rotate through 360

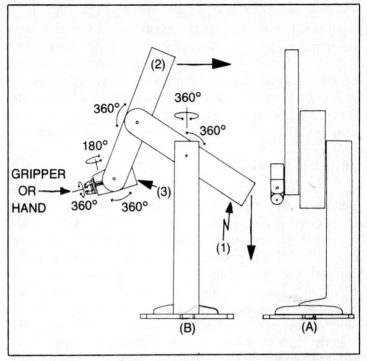

Fig. 1-4. The statics and dynamics of the Grivet robot (courtesy American Robot Corp.).

degrees around the base pedestal so that the gripper can move an object from here to there, in a sideways motion.

How can it be said that this mechanical arm emulates a human action? Let's take the situation where the arm, by definition, is to excel a human capability; do it better than a human can. To begin, assume the arm and the human are equal in speed, accuracy, and general operation. Time passes, the human tires, the machine does not. Now the machine excels the human operation. One can also visualize that the machine requires no breaks or time off for lunch and doesn't care whether there is light in the area or not (unless it is equipped with a visual type sensor). It doesn't care whether it is day or night. Thus, perhaps, to use the word emulate is to state the situation somewhat correctly. In this case emulate refers to only those actions that the arm is capable of performing. It obviously cannot duplicate *everything* a human arm is capable

11

of. This, of course, brings us to the type and kind of task where the human emulates the robot. The robot performs in a delightful manner as long as it is working properly and nothing goes wrong, and as long as all conditions of its task remain the same. But, unless it is an adaptive type robot, any change in the conditions of its job or task will not be accounted for until it has been re-programmed to compensate for those changes. The human, naturally, compensates automatically.

ARTIFICIAL INTELLIGENCE

One definition of artificial is that which is made by man to imitate nature. One can conclude therefore that artificial intelligence is a man-made imitation of nature. One can also assume that an imitation can never be equal to the real thing.

In robotics the actions of a machine may be of such a complex and delicate nature that one almost assumes the machine has a human-like mind direct its actions. Such machines have enough logical capabilities that what they are able to do seems uncanny. But if we fathom the human mind slightly further we find out that there is one aspect of the human mind which, so far as we now know, cannot be duplicated by a machine. That is the ability of the human mind to question. A machine analyzes, performs, has logic, and can make selections, but does it ever ask why?

Scientists are learning more about the human brain and its functions all the time. This marvelous electro-chemical device is said to be capable of doing billions of operations per second, of storing incredible amounts of information for extremely long periods of time, and of using such information in the solution of both new and old situations. It has been said that a human is the sum total of what he has learned—and he learns all the time. Since everything is fluid and changing, the human must constantly adapt and learn and adjust to meet changing conditions and situations.

So, some say that the only tasks and operations a human can perform are those that he or she learned to do previously, either by instruction, or by trial and error. Understanding this, one begins to think that a robot might be able to do as much as a human if it has a computer with a large enough memory bank to store all the information that comes to it in the form of learning (structured instruction) or in the form of feedback (trial and error). These trials are generated by some interior command arising from some need which its sensors can determine. The machine then senses a need, and looks through its memory banks for a solution. Finding none, it takes the closest approximation to a solution—some action which will cause the sensor information to change—tries it, and senses the result. If an improvement in the original sensed need results, the direction of the trial is correct. If no improvement results, or the need gets worse, then the direction of the trial is incorrect.

In this example we find the machine doing what many humans might do to solve a problem. But nowhere in this example is the concept of the machine stopping all operations and asking itself "Why is there a need of this kind?" This is what humans will do under similar circumstances. We do not know how to program a machine to ask that kind of question, exactly, although it may be approximated in complex programs.

WHO IS PROGRAMMED AND WHY?

Are we all programmed? That is a philosophical question which we, perhaps, might have to consider as we make mechanical machines more nearly like humans. We think of programming as a function related to computers and we think of the microcomputer being closely associated with robotics. These small, cheap computers have become the foundation behind the concept of robots or automated and intelligent machines. It is said that even computers are a form of robotics. They *do* have mechanical outputs (printers and graphing devices) as well as visual display outputs and many other output devices. What kinds of orders do they obey?

In the beginning such a machine is told to do task one and then do task two and from there it operates on certain conditions. For this example, the orders have been stated as follows.

1. Do f then g
2. If C then f else g
3. While C do f

What is implied, but not apparent, in using these orders is that the computer may be evaluating and scrutinizing masses of data which govern the use of these orders. It is far beyond the ability of the human mind to evaluate this data in the same systematic, logical, unemotional, and timely manner that the computer does in arriving at, say, an implementation of step 2 above. One might suspect that the human mind is programmed to evaluate certain amounts and types of data, and computers are programmed for another type. Human conclusions usually must be a consensus before the conclusions are accepted. We note that machines follow a man-specified routine to evaluate even the greatest amounts of data, but can only evaluate those types that man understands how *he* evaluates!

We can think through this problem by imagining that we are in a fast modern aircraft and suddenly, ahead, is another aircraft bearing down on us. Should we dive, climb, turn left, or turn right? What will the other pilot do? What if he maneuvers in such a way that the intercept possibility increases? It is possible that you have had a similar experience while you were walking down the sidewalk. Another human approached. You moved to your left, and he moved at the same time to his right. You moved to your right, and he mived to his left. There was no way you two could prevent bumping into each other. The consequences here were not disasterous; they would be disastrous in an airplane. Could a computer have solved the situation in either case? The FAA says yes. They have installed collision avoidance computers and control systems (robots, if you will), on board test aircraft for futher evaluation. The results look very promising. Collision avoidance systems are used on board ships at the present time.

What can the machine do that the human cannot do under these conditions? It has the ability to evaluate more data, faster, and evaluate it in a more complex routine than the human mind is capable of. For example, using radar sensors the computer can determine in an instant if the other aircraft is going up or down relative to your position. It can determine the speed of movement in all directions and predict or anticipate collision or no collision. The solution to the problem is then flashed to the human in a form he can most quickly and accurately assimilate—perhaps a light to indicate the direction he should steer his aircraft. The job is done well and safely by the two minds, electronic-mechanical and electrochemical, working together!

In a case where the human response time is too long, delayed, or cannot be accomplished in the time interval necessary, the secondary part of the machine operation—the robot steering—takes over and for a few precious seconds handles the aircraft. Aircraft autopilots, which steer aircraft along precision courses, at given altitudes, have long been recognized and accepted. They are being given additional duties all the time and tests have been conducted flying a passenger type aircraft from takeoff to touch-down without human hands ever touching the controls.

THE HUMAN MEMORY PROBLEM

The problem is that we forget. We also cannot assimilate much information at one time. It comes into our minds slowly, we retain it when we use it, and we lose if it we neglect using it for any length of time. Computers normally won't do that. Failure can mean the loss of memory in a computer, but normally the computer can accept, sort, and store all kinds of information quickly and for a very long period of time. Many computer memories today are magnetic tapes, magnetic bubbles, or various types of discs. Some are solid-state units which have information "burned" into them, so that when energized they immediately form voltages and currents which translate into data.

Sixty executives, all presidents of corporations were asked to select a number from zero to nine and they were asked to do this 100 times each. At the same time a computer was programmed to select, at random, a number from zero to nine, and it was permitted to do this 100 times. It was predicted that the presidents would choose the same number at least 10 percent of the time. That was not exactly correct.

They selected the same number 12.3 percent of the time; at least those who had increased their corporate profits during the previous operating years did this. Those presidents whose companies lost money selected the same number only 8.3 percent of the time. The computer selected the same number 10 percent of the time as predicted by laws of probability. This test was said to indicate that the successful executive had strong and accurate intuition and the results were reported in *Time* magazine. When these results were analyzed it was said that man's inefficiency, emotions, and the *nuances* with which a man reads data may inhibit his thinking or affect the accuracy of his conclusions. Using a computer to predict trends or to select courses of action based on evaluations of reams of data and then tempering the prediction with human judgment seems to be the best way to gain the utmost from the two worlds.

When we consider the world of advanced robotics, then, we must concern ourselves with the problem of how much intelligence we can build into a machine, how it will use that capability, and when we must insert ourselves into the control loops to achieve the best of the two worlds. We have to decide when the machine can function alone, and when it must be aided by human intelligence. Of course, it goes without saying that we will be forever trying to fabricate a complete mechanical-electronic-chemical mind which will permit humans to specify our commands and relax while everything is done for us! It has been shown by R.S. Aha of Grumman Aerospace that man's thinking process might be specified as shown in Fig. 1-5. In this block diagram form we can easily imagine how the brain of a very intelligent robot might function.

MORE ON THE GRIVET ROBOT

We learn still more from a deeper examination of the Grivet robot's characteristics. The arm sections as well as the whole arm itself are moved by direct current stepping motors which have built-in integral shaft encoders. This means that a motor can drive any arm section to within plus or minus .004 inch of a given position when the same command is

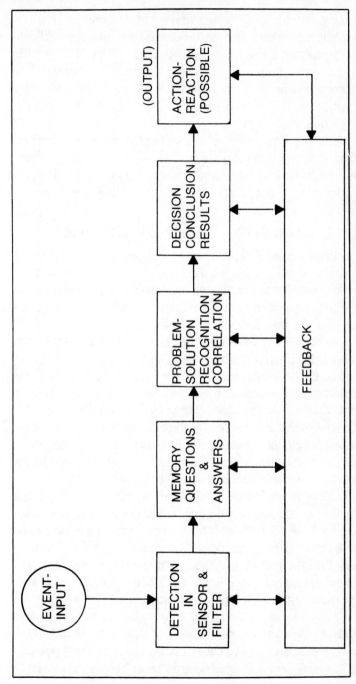

Fig. 1-5. One concept of the human thinking process.

re-issued. The motors are driven by a single-board microcomputer which can be connected to, and accepts data from, external lines. This computer has a built-in diagnostic capability to pinpoint machine-operational problems. The microcomputer uses a nonvolatile memory cartridge for program storage. Its controller is a 16-line, optically isolated, TTL-compatible unit.

Although the arm is electrically positioned, the gripper, or end-effector or hand, is pneumatically operated. The robot has a built-in air compressor to power this end unit. The use of pneumatics means very fast responses are possible.

ROLE OF MATHEMATICS IN ROBOT-ARM OPERATION

In an earlier discussion we mentioned the algorithm, or equation which a machine's computer solves to permit the robotic machine to function as the designers desired it should function. Considering the Grivet robot arm, we can gain a good insight into just what this means.

In Fig. 1-3 we examined the teaching pendant as the little box is called. On it we have seen that there are switches to control various movements of the various segments of the arm. We imagine, then, that when we want the arm to do a particular task, we just move the proper button and watch that segment of the arm move until it is in the proper position for the next segment to be moved. We next cause that segment to move until it reaches the proper position, and then we change switches to cause the next segment to move, and so on.

What if we make a mistake? If we are teaching the machine by using the little pendant to control the arm, what happens if we "overshoot" the desired position and have to do some manipulating and jockeying around to get the arm segment into the proper position? Will the robot arm then also jockey around each time it moves? What happens in this case? You can imagine that a person manipulating that arm for the first time would certainly not be able to handle the control pendant switches so perfectly that the arm would move exactly and precisely to each desired and necessary position for the completion of the required task. You merely need to try your hand at radio control of some small model with reasonably tight control to verify this idea.

So, what happens? You guessed it! Within a robot is placed a bit of artificial intelligence that is given the prime directive to assist both the human and the arm so that each, given its own faults, will not inhibit correct operation. The human fault is jockeying the controls when teaching. If the robot has a fault it would be its eagerness to respond, which it does with alacrity. So the computer's task is to eliminate the human fault and compensate for the robot fault to bring about a perfectly delightful operation which is pleasing to the human and maybe to the robot also!

Let us look at the computer and bring it under tight scrutiny for a moment. We know its job will be to solve equations and in this case they have to be equations of motion. We also know that when solving an equation of motion there must always be a reference point. In this case it is the base pedestal of the robot arm. Imagine that this pedestal is set into the floor, or mounted thereto in such a manner that it points north and thus the directions of south, east, and west are also defined. Another way of looking at this coordinate designation would be to say that north equals forward, south becomes backward, or reverse, east is right and west is left of the base structure. We can so define this arrangement if we imagine ourselves being located in the space occupied by the pedestal and have the specified body relationships.

From our location we look at the end point, or gripper, or tool to be positioned which is mechanically fastened to the end of the wrist. That is the point which we want to move to a given location in compliance with a series of commands. We will, of course, also want that end tool or gripper or whatever to do something at each end point. It might pick up an object, it might set down and release an object, it might position a welding element, it might operate a paint sprayer, it might draw a path from the beginning of movement to the end movement such as is done in graphics, it might wire-wrap a terminal or a series of terminals, etc. But the end point must move from somewhere to somewhere else very accurately and in such a manner that a duplication of its action will also be exact—perhaps for thousands of movement operations. Its repeatable positioning must be present and very accurate. Remember the Grivet operation is within plus or minus .004 inch!

So what does the computer do? The computer remembers the starting point and the end point of the gripper for each cycle of operation. A cycle will be defined here as that series of movements required to perform one task. The computer then computes the best path through three-dimensional space for the gripper to move to accomplish the task the human led it through so laboriously, and the resulting path equation or algorithm will be solved by the computer each time it runs the arm through that cycle. We thus learn that the gripper may not go through the same points in three-dimensional space that *we* put it through when we taught it where it was to move. It might move along an entirely different set of spatial coordinates to accomplish the same task better and more efficiently. It definitely will not remember any jockeying we might have done, or manipulating we might have had it do to get it where we humans desired it to be when it finally got there. All those little breaks will have been removed from the motion-path commands and the result, to everyone's delight, will be a smooth, direct operation.

WHAT IS A TASK?

We have used the word task and it may have a different meaning to different people. A task may mean one completed action, or one complete cycle. Here it means a movement of an arm-section point to one specified position in a series of end-point positions in order to accomplish a job. The end-point is the tip of the gripper or hand. Many tasks make a cycle. We consider one job accomplished when one cycle has been completed.

A vector is a line that has both direction and magnitude. In Fig. 1-6 we find three vectors with fixed lengths—V1, V2, and V3—and these represent the three arm segments of a robot such as the Grivet. They are attached to each other and to the base at one point as shown. Each arm segment can move independently. In this two-dimensional drawing all of

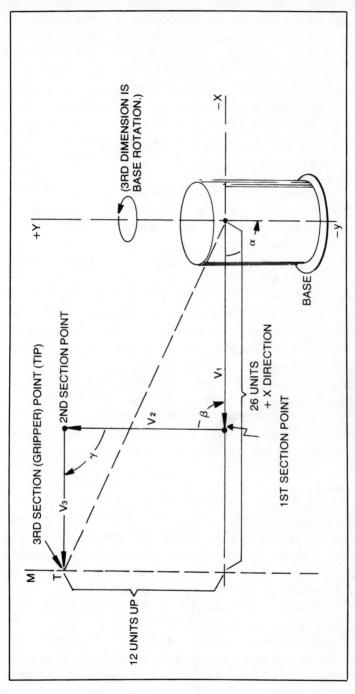

Fig. 1-6. Vectors describe the position of the 'end point' position of each arm segment.

the angles cannot be shown. V1 moves up or down by opening or closing the angle alpha (α). V2 moves left or right (as shown here) by opening or closing the angle beta (β). V3 moves up or down, and so moves its tip which we imagine to be the robot's gripper, by opening or closing the angle gamma (γ). With a little imagination you can visualize the simultaneous movement of all three vectors such that the end point might be moved up, down, left, right, or in a combination of these directions as required by the tasks. Of course, V3 could be longer so that if the angle gamma were reduced to zero, the end point would come around to the first section point and, equipment permitting, would coincide with that position. As shown, the end point has been moved up 12 units and out along the positive axis 26 units.

We might imagine that the end point is a welding flame or contact point for an electric welding unit. We might also imagine that we want the end point to move up from where it is shown to point above it, and that when doing this movement, the arc will be energized by another command so that welding would take place along the strip T—M. In your mind's eye, visualize how the vectors must move through their angles to keep that end point pressed against the metal being welded along the strip T—M. Angle beta will increase, as will angle gamma. It is possible, but not probable that the angle alpha would increase during this operation.

This is a two-dimensional drawing using just the up-down and left-right movements of the end point. In actual practice a robot arm will move in three dimensions, so that there would be a rotation of the vectors into and out from the paper also. This might happen if the end point had a gripper which was to pick up something from a conveyor belt on one side and move it to another conveyor belt located on the other side. The mathematics of the operation would then have to be such that the gripper might move down along the T—M line as shown, until the gripper tip was level with the origin point (0) or in direct contact with the X axis.

In the Grivet-type robot — and others of this type—the arm movement is specified to the computer in the form of the first section point, the second section point, and the third section or gripper point. When these points have been fed into the computer for all the tasks needed to complete the job, and in the proper sequence, the computer will solve the necessary trigonometric relationships for the best movement to reach each end point. If you need to refresh your memory concerning task and job definitions this is a good time to go back and do so.

GIVING THE ROBOT SOME INTELLIGENCE

We have indicated that a robot arm might be able to pick up objects from one conveyor belt and move them to another belt. There might be many objects on the first belt and the arm might be programmed just to pick up selected items and not all of them. There are two basic methods of accomplishing this. First, give the robot some eyes, and then match the shape of things on the belt against memorized shapes in the robot's memory. The memory bank here might be a cassette tape, so that if you wanted to change the shape of objects recognized, you would simply change the cassette software.

Another method might be by identification of the position of the object on the first conveyor belt. With a constant rate of movement by the conveyor, and precision dumping of objects onto its surface, the objects should come by the robot station in precisely separated distances. The arm could then be easily programmed to pick up anything (within its gripper capability) spaced that distance apart, move it, and return for the next object. A problem might arise if the object belt happened to slow down, the objects happened to be placed on the belt incorrectly so the gripper could not easily grasp them, the conveyor belt happened to speed up for some reason.

One use of this idea is the welding of automobile bodies by robots. The bodies come by at precise speeds and precise positions. The arms move to precise positions also and weld and weld and weld. . . and do a good job of it! In robot systems where some intelligence in the form of contact feedback is incorporated, if the body happens to be just a little out of line, the arm will move to compensate for this discrepancy, also a speed monitoring system on the conveyor track or whatever, will keep the robot system informed so that even if the line speeds up or slows down, the arm will adjust to compensate for this change. No doubt you have thought of other ways in which the robot might sense or determine or find objects on that first conveyor belt. Some other means which have been considered are: temperature of the body, size of the body, and—believe it or not—actual recognition of the body even if it happens to be in some unusual orientation on the belt! This latter case is very important because it means that such robots can actually detect and pick out specified objects among many other objects on a big tray or on the belt surface. This is important in some assembly-type operations.

Giving the robot intelligence then means that it will have a computer to solve motion movements, it will have sensors to assist it in accomplishing its job, and it will have some kind of anticipation and adjustment circuit in case its program doesn't exactly fit every situation.

In many current situations where robot arms, or robots with fixed locations, are used and there are a multitude of robots "employed" on a line, it is possible to have one intelligent robot controlling the operations of many other dumb robots. The tasks may be similar, and the jobs may be the same. In some cases the intelligent robot may control another robot in such a manner that the second robot does some operations or jobs which assist or complete the job the first robot is performing. This is the case when one robot with one arm actually needs two arms to do the job. The solution is to get a second one-armed robot and program it from the first one so that the necessary actions are then accomplished.

Another form of intelligence built into a robot such as the Grivet is that of delay. This means, in human terms, the programming necessary to cause the arm to wait for an object to arrive on the conveyor belt, if it has not arrived when the arm is moving to pick it up. Robot arms can move very fast. It may be necessary to tell the robot to move its arm into a pickup position and then wait a specified time before moving to actually pick up the object. Suppose a robot arm puts on a bottle cap and then has to screw it down on the bottle. That takes time. The arm must not move the bottle until the screwing operation has been completed.

THE ROBOT DIAGNOSTICS

It has been said of an intelligent robot that it, like its human counterpart, is always checking itself to see if it is okay. A human begins to moan when it gets a sore throat, or when muscles ache, or when the stomach is upset. A robot moans also, but in a different tone, as its sound comes from a bell, or siren, or screeching of gears, or clang of parts. Both the machine and the human must have some means of diagnostics to find out what the cause of the problem may be.

With humans there is conversation and a probe into "memory banks" to find related experiences to give clues to the trouble. Second there is a probe of memory banks to find out what to do in case the problem justifies a more serious approach than an aspirin, bed rest, or other nice homey remedies. Probably the diagnostics will start with a temperature reading, then pulse reading, then X rays, and so on. Of course it will involve a probe of the memory banks of some good physician who will try to relate the problem to something in his experience. He will then probe for the cure used or effective treatment to reduce or eliminate the problem.

With robots the approach is much the same. Within a good robot system there will be sensors to monitor the various moving parts. Simple remedies will be found in the minds of attending humans once the robot has voiced its complaint by means of its sensors and indicators. Within the intelligent robot's brain will be found memory banks that know what the motor speeds should be, when a motor or other shaft has cracked or broken, when there is an oil leak or pressure leak in a flexible line, or that there is no voltage or incorrect voltage to some critical point in the system. The memory banks recognize these problems because they have been programmed with the correct operational conditions, and thus, by comparison they can determine when something is wrong.

Some robot systems constantly monitor all parts of their systems just to make certain that a failure does not occur. When the machine senses that something is going wrong, it may alert its human operator and say that it needs maintenance, or it may shut itself down while keeping its indicators on to tell its human companion where to look and what to do to get it going again. These kinds of systems run a diagnostic check on themselves when they are first energized to see that everything is all right.

In a previous paragraph we discussed the need for a test system to test the tester, etc. If you combine that knowledge with this series of paragraphs you can determine the difference in requirements. Here the memory banks of the computers have the necessary knowledge to know when an output is correct or not. If the monitoring feedback, for example, does not correlate with what is in the memory banks, either because the memory bank information has changed or the monitoring information has exceeded specified limits, then the system shuts down or reduces operation to a safe value. Meantime there is an automatic alarm to the human supervisor to do some checking and find out what the problem might be. Alarms might be visual, in a high sound-density situation, or even be a remote paging system in case the human is not around. With the modern technological advancement in synthetic speech capability, the machine might just start speaking to its human supervisor: "I've got a pain in my tentacle!"

ROBOT COMMUNICATION

Let's expand voice communication slightly and think about robots talking to robots. Some people write about the danger of having robots communicate with each other. One writer says, "Many Americans fear this new age of computers that talk to other computers and operate machines. Technology is moving so fast, in so many areas, that people are afraid of it because they are not familiar with it."

In assembly line operations where many robots do the same type of job, it is useful for one machine to direct the work of other machines. The Grivet series, for example is so designed that one might have four slaves and one master or supervisor for these four units. It is said that the operation of many such machines is "orchestrated" so that they can perform process control, handle inventories, handle outputs and inputs to stockage areas, and control the flow of items in batch lots on various types of moving tracks, belts, or whatever.

Ask yourself what that word orchestrated means? If we devote some thought to it, we realize that in a large orchestra there are many different instruments, different parts to the melody and harmonic effects which all blend together in order to produce the final, ear pleasing sound. Thus we arrive at the conclusion that orchestrating a group of robots means the blending together of many different processes, automatically performed in a timely, coordinated manner to bring about the fabrication of something, or the completion of some process comprised of many parts. It could, of course, also mean the systematic and coordinated demolition of something if that were the process desired! Robots take many forms even when they work singly as shown in Fig. 1-7.

Yes, robots on assembly lines talk to one another. They may communicate that they have completed a task, they may ask for help from a nearby machine so that two arms can be brought to bear on a process instead of the one with which each robot is normally equipped. They may govern the flow of materials or regulate the inspection and processing of the materials as required. We can imagine one robot at the end of an assembly line, trying to fit some parts together and finding it difficult, communicating via its own computer and connecting lines back to the robot which is responsible for machining the parts: "Hey Buster! Get on the ball down there! Reduce your product by 1/10,000,000 inch!" We can also imagine an old robot instructing a new one in the tasks it must perform.

One big problem in manufacturing is that there are many small and different batches of parts and items which must be made. This means, of course, that the robot assembly line must change its actions and operations whenever a new batch is to be handled. Consider an assembly line of up to 30 robots in length. What a way to get into a new dimensional system—30 robots in length! Normally, people would simply use new approaches and methods to handle the new and different batches as they come down the line. What do the robots do?

If there is a change in products and there is good and orchestrated communication between and among the robots concerned, a human merely inserts a new program cartridge into the master computer. The human changes the memory for just one robot in this system and does not have to change the memory of all the robots as he or she might have to do if they were operating as individual units. It *is* advisable to have good, complete communication among the robots in manufacturing or processing plants!

People who fear these connections, will learn that the communication is not intelligent reasoning, as sometimes takes place when humans communicate. Among robots, the communication consists of a series of drive and feedback signals, which cause the robot receiving the information to do something. When it does what it is supposed to do the reverse communication takes place which informs the master that it did what it was supposed to do, and to what degree of precision it did it. If the reverse communication doesn't take place, the master knows the slave isn't working properly.

COSTS

Of course robots cost money. They start at $10,000.00 per unit for manufacturing types, and $3,000 for some of the android types we will discuss in another chapter in this book. Android types are hobbyist's toys right now, but imagine that someday they will be household servants and let us all live life in Utopia!

But, what about costs in the manufacturing situation? One might evaluate the cost of programming and operating a robot vs the cost of using human hands to do the job in cost per hour. Remember that the robot doesn't complain, doesn't worry, doesn't take rest breaks or lunch breaks, is always on time, and can work 24 hours a day without getting tired (worn-out maybe, but not tired). Each robot has an operating cost per hour which can be compared to the human cost per hour and when the purchase has been amortized the relative costs of the two will show something interesting. Although the robot won't ask for a raise, its operating costs will increase somewhat simply because the costs of energy will increase, and it uses energy. It doesn't get tired, but it should have some kind of normal maintenance routine and humans do not normally have to be maintained on the job. It is true that

Fig. 1-7. Prab robot unloads die casting machine at Halophane (courtesy Prab Conveyors).

the robot machines will do lots of jobs humans don't like to do and it will do them well. This will release humans to do other, more profitable work. So, if humans are to become future Robot Masters then they must prepare themselves now for profitable and rewarding jobs. Technology will not stand still.

FEEDBACK ACCURACY

We show, in Fig. 1-8, a kind of illustration which may indicate how the accuracy of measurement of the end device position may affect the general operation of the robot as a whole.

Along the X axis we have the physical position which can be measured by the feedback potentiometer, ac wave comparison unit, pressure feedback, or whatever. Along the Y axis we have the accuracy of repeatable operations which are performed time after time after time. Note that we never reach a perfect operational state. If nothing else, Murphy's Laws will prevent that from happening. But we do reach a close proximity to the state of blissful perfection, providing that the feedback measuring elements each can provide the high degree of accuracy required.

We have shown a linear drop off of accuracy. This may not be exactly correct. Each machine may have a different curve even though each machine may be constructed exactly the same. The operational environment, the type of load, the use to which the robot machine is put may well affect its repeatable accuracy. From this illustration, then, we begin to get some understanding of the complexity needed and the precision required in order to exactly locate and assemble small batch parts, or to perform operations where a few millimeters of movement may be the whole ball park. We are reminded that difficult measurements are not exactly impossible. Back in the Dark Ages of the 50's when the atomic bomb was being developed, there came a requirement to measure extremely small amounts of U-235. Using a delicate, and almost unthinkably precise, balance the scientists concerned were able to measure U-235 in amounts of millionths of an ounce. Of course you would not even dream that such a measuring unit might be used in other than a laboratory

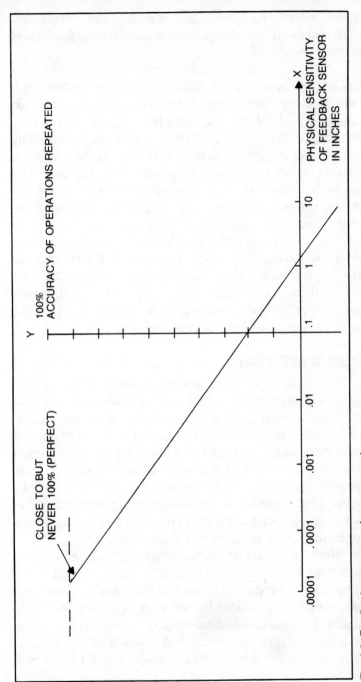

Fig. 1-8. Feedback accuracy and measurement.

operation—or would you? Think what that kind of measurement possibility would mean in terms of precision feedback of an advanced robot!

Now we are imagining the use of a machine robot to do jobs that are being performed by humans using microscopes. This requires them to move with slow, delicate precision because the slightest error could be costly or irreversible. We imagine that if such precision could be accomplished by robots the costs of many items of modern technology might be reduced and their reliability might be vastly increased. The key to this precision is the ability of the machine to function on the smallest of signals, and to physically measure the smallest value we can think of.

This leads to another problem area—maybe it's Murphy's Laws again. The smaller the signal or the more sensitive the feedback, the greater the danger from extraneous electrical and mechanical noise. Noise means error signals and noise is the enemy which must be overcome in the fight for greater precision and better computation.

SPEED OF OPERATION

In a study of servomechanisms, which form the basis for all robotic operations, we find that the greater the speed of movement the harder it is to prevent overshoot and oscillation. There are, of course, mathematical solutions to many of these kinds of problem, and they are used. We want to plan beyond the present day capability and consider what the truly maximum speed of operation of a robot system or an individual robot might be. We are considering power behind the movement, and such amounts of power that the overshoot and oscillation problems again rear their ugly heads.

How fast is fast? Why do we need to increase the speed of operation? Is it useful, or required, or just a desirable development? Is productivity, and therefore profits, dependent upon speed of operation? Do we want to work toward that ultimate blinding speed of assembly, or are we going as fast as we can go considering our knowledge, and capability of making advanced robot machines. Consider the CYRO 5 + 2 advanced robot made by Advanced Robotics Corporation. This machine has two coordinated arms instead of one to accomplish welding operations. One hand has a five-axis torch motion (x,y,z,c,a,) and the second hand has two-axis standard motion. Both hands are said to move smoothly and accurately to accomplish many tasks. Of course the coordinated motions must be computer directed or program directed.

Computers are being improved every day and their speed of operation, output capability, and controllability increase every minute. It seems that we must expect physical motions from our Advanced Robot machines to keep pace, and these motions, at vastly increased speeds, must be accomplished without overshoot or oscillations. Also required will be the ability to perform, for long periods of time, these fast and precise operations, i.e. a high degree of reliability and very low down time ratios. In Fig. 1-9 we illustrate the CYRO 5 + 2 type robot and some motions and jobs associated with it.

It is interesting that this robot is said to move in *rectilinear coordinates*. One must envision that this is accomplished through computerization of the angular type coordinates and movements which evolve from the movement of the end device shown at B. It is said that rectilinear type construction provides a stiffer structure for greater accuracy and smoothness than would a comparable angular movement type. Also, it is claimed that a rectilinear system provides greater safety since people tend to think in rectilinear coordinates and thus can anticipate the machines motions.

If one examines Fig. 1-9 (A) it is easy to see what is meant by the rectilinear motion. The arm moves up and down, in and out, and its support column can move left and right. There is no turning of the arm or its support column elements. There are rotary motions associated with the end gripper or, in this case, the welder unit. No doubt some will say that there are certain advantages for the rotary movement systems. That, of course, will be a function of what kind of machine is being manufactured and its particular application. I

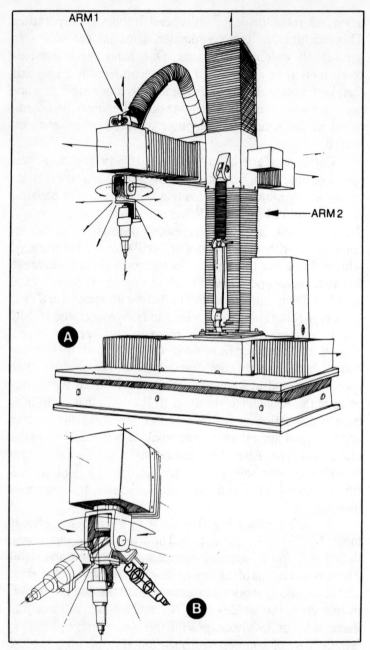

Fig. 1-9. (A) Cyro 5-plus-2 two arm advanced robot. (B) Movement of "end unit" of Cryo 5-plus-2 (courtesy Advanced Robotics Corp.).

ARM 1

ARM 2

A

B

Fig.1-10. Cyro 5-plus welding (courtesy of Advanced Robotics Corp.).

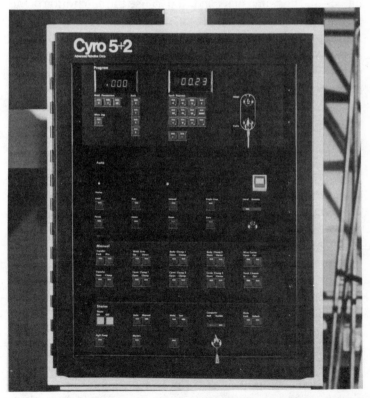

Fig. 1-11. Cryo 5-plus-2 control panel (courtesy Advanced Robotics Corp.).

would suspect that there is a use for each type of machine, and that usage may govern what type advanced robot is procured by a manufacturer.

SAFETY

In advanced robotics we must always be conscious of safety just as we would be conscious of this requirement in any plant or operation where machines are used. For the advanced type robots we have so far considered, we recall that they are usually fixed in place, that they have a certain radius of operation, or space volume in which their actions are confined, and thus, regardless of how they move or what they do, if we restrict ourselves from entering that action space around each robot, we should be safe enough.

If there is a malfunction which should, by some very unique and very unusual condition, cause the separation of the physical parts of a robot from its base structure, then, of course, that might make unsafe any other volume of space around the unit. But this is a case which is so rare that one might consider it just doesn't happen. Normal safety procedures such as are well developed and practiced in machine-containing plants and operations, will normally suffice, and will protect those humans involved when they are in the same physical areas as the advanced type robotic machines. One would wear protective glasses if the machine is producing particles of any type which might be injurious, or if the rays from the operation (welding) might be injurious to the eyes. Clothing must be such that sparks or whatever type of remnants might be cast off from the tasks performed will not ignite or penetrate or cause other troubles to the human contained therein. Proper footwear would be very nice to have over your toes just in case you are close and the gripper opens prematurely, dropping a large object to the floor. Normally, however, one would be walking or spending much time near the robot assemblies. Observation is through instruments and TV type tubes and visual inspection from isolated and protected control booths. Only in the case of trouble with a particular machine is human presence required in its particular space. Then adequate precautions are taken when initiating the repairs and that will serve until that day when the advanced type robots repair themselves!

The mobile type robot may present another situation. Its volume of active space may also be confined but that might be a large space and a moving space and one must then practice safety as one does in the street with moving automobiles! We can just imagine our reaction if we are in a large building and suddenly, silently, this monster machine comes bearing down on us with its four arms waving madly. The arms are moving—not because it is angry, for robots don't get that way—because that is the way advanced robots normally move! Of course we could be on the regular robot track as defined by some invisible substance on the floor and it may be just moving from one job location to another in accord with some of its prime directives. Yes, unless it is programmed to stop for obstructions it might well run over us! But even that, somehow seems remote as a possibility. Certainly, any advanced robot of the mobile class will have such sensors that it will stop and inquire—perhaps verbally—if it encounters anything unusual in its path, or on its job, or during its performance of its normally scheduled and programmed tasks.

ROBOTIC SPEECH

Not many years ago it was considered almost impossible to make a machine which could actually talk back to its operator. As we well know, that impossibility has been replaced by fact. One of the companies in the forefront of speech synthesis in Texas Instruments, whose *Speak and Spell* solid state learning aid has started the machine talking revolution. Figure 1-12 shows the chip which resulted from many years of research and development in the speech producing effort.

The heart of TI's Solid State Speech systems is this little monolithic speech synthesis chip which was invented by TI. It actually generates electronic signals which, when reproduced on a loudspeaker system, sound like the human voice. The chip models the characteristics of the human vocal tract.

Machines that speak, either to acknowledge or convey information, are deemed the next area of the technological

Fig. 1-12. Voice synthesizer chip (courtesy Texas Instruments).

revolution according to TI. They are engaging in a big program of research and development related to computers, industrial machines, telecommunications, automobiles, and to the entertainment market. Now even an automobile, properly equipped with a small diagnostic computer, will tell the mechanic what is wrong or what needs adjustment in or on an engine. How nice that concept is for the backyard mechanics! If the machines will only explain carefully and in detail exactly how to correct the problem, as well as telling you what the problem is, then we've got it made! That you will find more and more speaking machines in every phase of our lifestyle is almost guaranteed by such indicators as TI's separate speech organization to serve as a focal point for all their new applications and developments in this field. Let's take a closer look at that chip by examining Fig. 1-13. You can amaze yourself by examining Fig. 1-12 and locating the Pipeline Multiplier chip and then looking at Fig. 1-13 to see what is inside that little device! But as you see from Fig. 1-13, that's only a part of the circuitry involved.

TI defines synthetic speech as either a word, phrase, or sentence, or a complete or unique sound. A lot depends on its

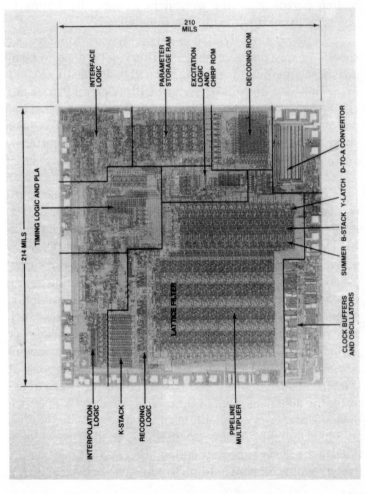

Fig. 1-13. Geometry of the TI speech chip (courtesy Texas Instruments).

210 MILS

214 MILS

INTERFACE LOGIC

PARAMETER STORAGE RAM

EXCITATION LOGIC AND CHIRP ROM

DECODING ROM

TIMING LOGIC AND PLA

D-TO-A CONVERTOR

SUMMER   B-STACK   Y-LATCH

CLOCK BUFFERS AND OSCILLATORS

LATTICE FILTER

INTERPOLATION LOGIC

K-STACK

RECODING LOGIC

PIPELINE MULTIPLIER

39

use, duration, and application. A word, for example, is defined as a second of "utterances", and only TI can explain what is meant by that! Perhaps that is related to the 100 word vocabulary of the chip in Fig. 1-12, which can be spoken or uttered in 100 seconds. Of course improvements in vocabulary, and if required, speed of making the sounds, will be produced as the need arises. We are truly approaching that day when all commands and instructions to our advanced robots will be given by simple speech commands, and the robots will also advise us of their condition and when they need maintenance they will provide us with any other information pertinent to the overall job, simply by telling us verbally, or vocally, or making some sounds!

As pointed out by TI it is relatively easy to increase a robot's or machine's vocabulary. This can be done in many instances simply by appending a sound to an existing sound. In the polled-status mode, the host CPU (Central Processing Unit) issues the address of a word, sets the talk command, and polls this status bit to determine that the word has been spoken. Delays between words and sentences are inserted by addressing the particular delay word which is processed as if it was just another word.

The monolithic speech synthesis chip uses Linear Predictive Coding (LPC), which duplicates the human vocal tract. As this name implies, LPC is a linear equation which formulates a mathematical model of the human vocal tract. Thus it is possible to predict a speech sample based on previous samples. LPC is a technique of analyzing and synthesizing human speech by determining from the original speech a description of the time varying pitch and energy using digital filters which also reproduce human sounds when excited by random or periodic inputs. Because digital impulses themselves cannot be used to drive loudspeakers, it is necessary to have, on that same chip, a Digital-to-Analog (D/A) converter which transforms the digital information into signals required to energize the loudspeaker or earphone. On the chip shown in Fig. 1-12, an 8-bit Digital-to-Analog Converter is used. This is identified in the lower right hand corner of Fig. 1-13. Codes for the twelve synthesis parameters (10 filter coefficients, 1 pitch, and 1 energy) serve as inputs to the synthesizer chip. These codes may be stored in a ROM (Read Only Memory). When the codes are decoded by on-chip circuitry (Fig. 1-13) they produce the time varying signals descriptive of the LPC model, or human voice sounds.

The input to the digital filter takes two forms, as we have stated. They may be periodic or random. The periodic input is used to reproduce voiced sounds that have a definite pitch such as vowels or voiced fricatives (sounds formed and pronounced by forcing the breath through a narrow opening between the teeth, lips, etc, such as f, s, v, and z). In the TI the fricatives are z, b, or d. A random input to the chip models unvoiced sounds such as s, f, t and sh. The speech synthesis chip has two separate logic blocks which generate the voiced and unvoiced sound patterns. The output from the digital filter, drives the Digital-to-Analog Converter and *that* drives an amplifier which drives the speaker. The rest of the chip, as shown in Fig. 1-13, consists of the integrated array multiplier the advanced 10-stage lattice filter, the adder and multiplier, and the delay circuits. A complex bit of circuitry, but since it is all on a chip it is easy to obtain and use.

A final note about the speed of operation. With the unit shown (and newer units will have a higher speed of operation) the rate is about two inputs per each five microseconds. TI says it takes twenty multiply and accumulate operations to generate each speech sample, but the circuits can still generate up to 10,000 speech samples per second!

SUMMARY

Throughout this chapter I have presented a glimpse of what the world of robotics consists of. We have found that it is a vast world indeed, encompassing mechanical engineering, electronic-electrical engineering, light, sound, chemistry, atomic and computer engineering. Some disciplines which we have not mentioned may be found in this field. So we learn that it is a big area of study, design, and development. We have called advanced robotics a science and an art and we have presented a simple type of definition: "The design, use, and operation of machines, which are computer controlled to do human desired tasks." We also indicated that within the control framework might be a human brain as a part of the computing system.

There is an organization, dedicated to manufacturing type robotics which has given a somewhat different type definition of a robot. The Robot Institute of America has come up with a typical engineering type definition: A reprogrammable, multifunction manipulator designed to move material, parts, tools, or specialized devices, through various programmed motions to perform a variety of tasks. The manipulator in this case is what we have called the end product (or tool) holder, or gripper.

While it is true that robots of any type, industrial or not, will get smarter as time progresses and will be able, through use of many sensors which humans do not have, to accomplish many tasks relating to jobs we have for them, in a most pleasing manner, they still must be programmed by someone. The user may not develop the software necessary, but someone has to develop it, and someone has to develop the command instructions so that an android robot, for example, will respond to your spoken commands.

Webster's dictionary has another definition of a robot, and from this can be derived another definition of the world of robotics: "A robot is a machine in the form of a human being that performs the mechanical functions of a human being." Of course, this has been the general concept carried about in many human minds for a long time, but it is not necessarily a true definition as we apply it to today's electromechanical marvels. The Madison Avenue personnel devote hours and hours to research and planning and burn much midnight oil to prepare sales pitches which involve the magic word "robot(s)", and in many instances what they define as a robot, for sales purposes, is not a robot at all. We have to make allowances for the kind of sales personnel who jump on the bandwagon, and present things which are not exactly correct, according to the definitions presented herein.

It is very interesting that the Lord Company of Erie, Pennsylvania hopes to market, a robot hand made out of a kind of sponge and filled with a grid of wires and sensors which are so arranged that it will have an almost human sense of touch and feel! Robots are here now! Let's move ahead in our study of some advanced types, and some hobby types.

The Domesticated Android

While industry plans and researches and builds toward an automated improvement in their operations, there are some companies who are looking into the possibilities associated with the so called domesticated robots. These are special household and office machines which, they say, will become as familiar in the future as a household pet, and they will do much more than any pet—even though it be well trained.

We now examine some types of domestic robots as they appear today in various market places. Through the courtesy of Quasar Industries, we show an example of what a domesticated android might look like in Fig. 2-1. Particularly interesting is the robot's somewhat formidable vision window, which seems somehow to conceal things that we really shouldn't know about. In the market place, trade fairs, on TV, and on stage robots of the type shown are being seen more and more often, and their capabilities are being enhanced every year so that they are approaching the state where they might be called android instead of robot.

THE ANDROID EXPLAINED

What is an android? In the field of robotics there is a good and precise definition.

> *android*—A human appearing type of machine which duplicates in appearance and some actions the human body and its functions. May converse, and does respond to external stimulii with pre-programmed responses. This is the type envisioned by sci-fi writers as the companion to the hero in far-out escapades.

I add, in this writing, that the conversation capability is now enhanced by a capability to formulate responses by analyzing the questions asked, and thus it does not depend solely on programmed responses as was the case in the past. Also we must now add that the android is capable of sensing and analyzing to a much more advanced degree than it was in the past.

So what is the definition of a robot? A natural follow up question.

> *robot*—A general term which stands for an automated unit which does have some kind of human symbolization in its operation or physical structure. It may not look human, but it seems to be able to perform human type functions and actions.

In this book we shall find that there are many industrial robots in operation and that they do, indeed, perform many human-type mechanical functions. Usually we will find that they are programmed in some manner to perform the tasks that they have to accomplish to do a job. We also note here that the most advanced type of electrochemical-mechanical machine it is possible to conceive of has been defined in these same references as a cybert. So, while we consider the word android because it is well known, and conjures up visions of C3PO in *Star Wars*, when we consider household tasks we really have to think of a cybert. We will also use the word robot because it is commonly used, and is understood to represent a type of machine which can do things, as indicated in our definition.

We find that one of the tasks confronting us at this moment is sorting out what kind of machine we are to be concerned with. The machine gets a different name each time its intelligence and mechanical ability advance, it seems. In

Fig. 2-1. A domesticated android (on the right) (courtesy Quasar Industries).

order not to get lost in semantics, I suggest that you choose a name which has meaning to you and use that. What we will be considering is a *thing* like that in Fig. 2-1!

SOME MECHANICAL CONSIDERATIONS

Examine Fig. 2-2 for some of the mechanical details of this kind of robot. Here we find something concerning the dimensions and sizes of the various external parts of the body. We also learn something of the movement of this type robot. Of course these considerations can be changed with some

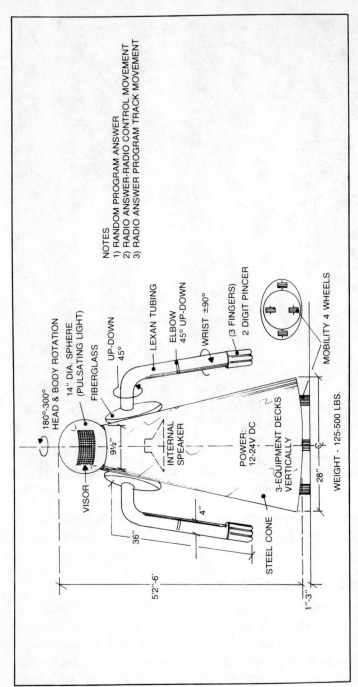

NOTES
1) RANDOM PROGRAM ANSWER
2) RADIO ANSWER-RADIO CONTROL MOVEMENT
3) RADIO ANSWER PROGRAM TRACK MOVEMENT

180°-300°
HEAD & BODY ROTATION

14" DIA. SPHERE
(PULSATING LIGHT)
FIBERGLASS

UP-DOWN
45°

LEXAN TUBING

ELBOW
45° UP-DOWN

WRIST ±90°

(3 FINGERS)
2 DIGIT PINCER

VISOR

INTERNAL
SPEAKER

POWER:
12-24V DC

3-EQUIPMENT DECKS
VERTICALLY

STEEL CONE

9½"

MOBILITY 4 WHEELS

28"

WEIGHT - 125-500 LBS.

36"

4"

5'2"-6'

1"-3"

Fig. 2-2. Mechanical details of one type robot.

48

adjustment of the drive motors and limit switches and control mechanisms. It is to be noted that movement is accomplished by electrical signals to two motors, one on each side of the body, with two more wheels—fore and aft—used as stabilizers. One would suspect that the drive motors in some applications would be stepping motors driven by a series of pulses. This permits very precise control. When the motors are in step together and in sync the robot goes straight, and turning capability comes when one motor is stepped faster than the other. The amount of stepping difference governing the speed of angular rotation, or turning.

The power supply would, in any type of autonomous robot or android, be a battery. One would desire a light weight, long life, easily and quickly re-chargeable, high ampere capacity unit. Of the types common at present the nickel cadmium or silver-zinc batteries come closest to meeting these ideals. Other types will be an outgrowth of developments for the space exploration programs, and for the expected electric automobiles which so many manufacturers are now researching. An ideal robot will have its own charging unit built-in and the unit will be able to find a wall socket itself and govern its own charging operation. Some robots have been built which can do this at the present date.

WHAT WILL A DOMESTIC ANDROID DO?

One of the questions posed is "What will it do?" If one is planning beyond just the construction of a robot, then one must plan tasks for it and determine how the robot will perform those tasks. Here is a list of what Quasar Industries believes a domestic android can be programmed to do:

●Answer the door when guests arrive. Take their garments, place them, and announce the visitors.

●Using a standard tray, serve pre-mixed drinks hors d'oeuvres, and perhaps even meals, provided someone replenishes the tray.

●Using several accessories, the robot will, vacuum rugs, polish floors, and dust general living areas, being guided by a computer program.

●Monitor the average home or apartment for fire, smoke, or unauthorized entry during any specified time period.

●Constantly monitor the house temperature and adjust by direct control (mechanical) or by remote control (sending signals to another unit).

●Monitor children during play or rest-time, or monitor convalescing persons sounding an alarm if the normal activities change significantly.

●Respond verbally to verbal commands and questions and provide information which is stored in its memory banks.

●Help provide education, amusement, and display cultural information.

●Cut the grass, rake the yard, take the dog for a walk, and other outdoor activities.

The ideas associated with "What will it do?" are not at the zero level. There are more ideas in this area of thinking than there are dollars to develop them. This means that the most important tasks will be automated first and then the smaller or less useful tasks and jobs will be accomplished later. It could be interesting for you to sit down and with a pencil and paper make a priority list of the tasks and jobs around your home, carrying this list out as far as you can possibly vision the work to be done.

Quasar gives us a titilating thought in an example of a homemaker who is getting ready to go to work. Having decided that the rug needs vacuuming, the command unit is taken up and the homemaker commands the domestic android quickly with the date, time, and function code. The robot then sends a confirming reply. Even the concept of having a robot prepare meals is not really far-fetched. It might be if the meals were prepared in the manner shown in Fig. 2-3.

However much amazement might be provoked by having such a robot prepare meals, there is a certain danger when such a machine fools around with flames, heat, or microwaves. The microwaves are not supposed to cause interference, but they might!

One suspects that the meal-preparing robot chef will not be a machine as shown in Fig. 2-3, but will be a built-in expansion of the time-cooking oven which is currently on the market. It would seem that we already have a computerized cook, what we need is a mechanical preparer and a mechanical device to get those dishes into a really good dishwasher. So, perhaps a combination of machines is needed in the kitchen. One, a robot to get things from cold storage and prepare the dishes, another machine to cook them, and finally, the robot to serve the meal, and place the dishes in the dishwasher. The automatic garbage disposal for everything, papers, broken dishes, bent utensils or anything, must also be present and in operation to clean up the rest of the mess. We offer a diagram in Fig. 2-4. As shown, you would simply tell the input microphone what the menu for the three daily meals is to be, state the time the meal is to be served, and note that all ingredients are on file. The computer takes over and directs the conveyors and ovens and stoves and robots to the end that the meals are prepared and delivered to the table at the times specified. Of course the human responsibility will be to insure that the bins and refrigerator stores are kept full and in proper placement. Eventually one might be able to eliminate human errors in this job by getting food delivered in response to computer instructions and sent directly to the food bins or freezers, automatically. Thus, once the meals are planned for the week, month, or year, the time frame being of your own choosing, all one has to do is to be there on time and eat!

We see the use of mechanical-electrical androids or robots in this example. Let us consider the same job from a slightly more serious perspective. We then ask ourselves if it is possible to completely automate a kitchen. The vision of a completely automated kitchen does not have robots or androids. It will have a multitude of machines and conveyors and arms such as the ones we have shown in this book. The "pick and place" one-arm industrial robots can do the task of picking and placing various food items, mixing and combining ingredients, and selecting various tools needed for those various chores. Fixed-location two-handed robots are currently being used in some factories for various types of assembly and some machines have vision sufficiently well developed to be able to pick out various things from other groups or items, so why

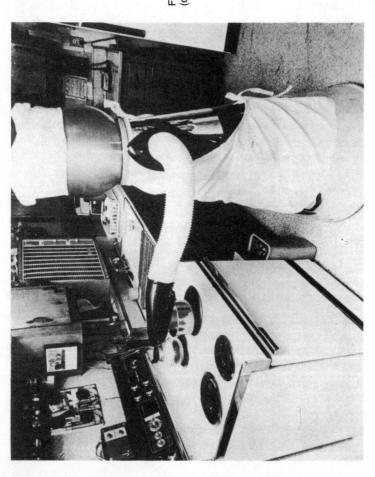

Fig. 2-3. The robot chef in action (courtesy Quasar Industries).

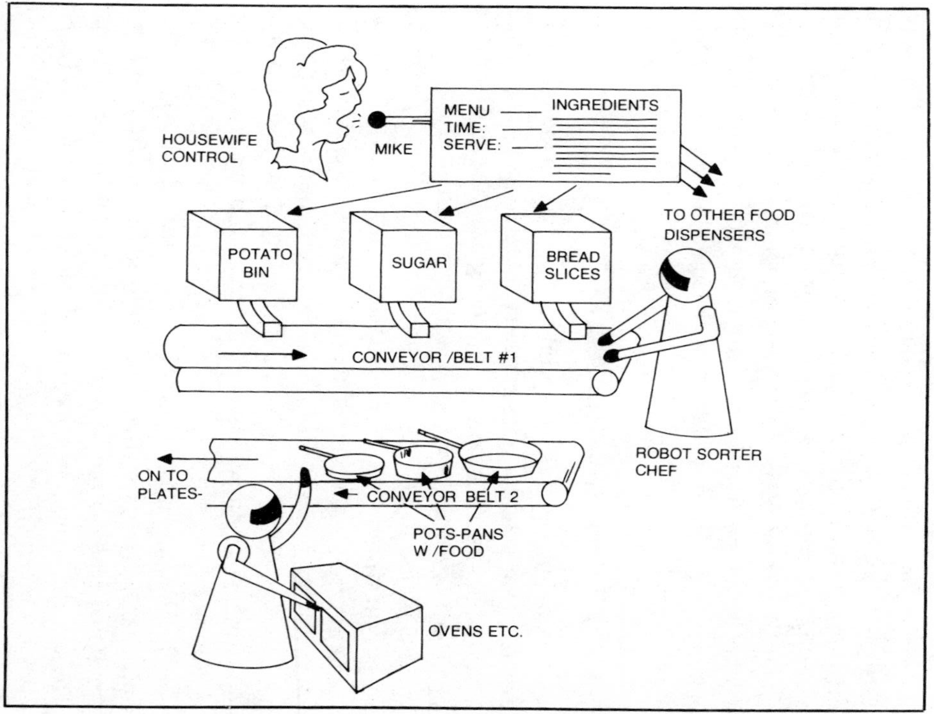

Fig. 2-4. The robot-controlled kitchen.

HOUSEWIFE CONTROL

MIKE

MENU
TIME:
SERVE:
INGREDIENTS

TO OTHER FOOD DISPENSERS

POTATO BIN

SUGAR

BREAD SLICES

CONVEYOR /BELT #1

ROBOT SORTER CHEF

POTS-PANS W /FOOD

CONVEYOR BELT 2

ON TO PLATES-

OVENS ETC.

53

isn't it likely that food could be an item from which selection must be made.

Let us consider some types of robots currently developed and learn what they can do. The University of Rhode Island engineers have developed Mark IV, a robot that can select parts from a jumble of parts in a bin. Some people at California's Stamford Research Institute have developed a seeing robot arm which can select the proper tool from many tools on a bench and use that tool to bolt down the cover on a small engine. At Carnegie-Mellon University they are working to enable a robot's brain to accept orders spoken in the human voice and, they say, the machine can now distinguish among some 1,000 words spoken to it, and it will ignore strange type human sounds which are not words. Joseph Engelberger of Unimation believes that the future world of humanistics will be laced with robots of the HAL type, and this, you will recall, is the type robot which doesn't appear as a robot at all, but is a space station controlled by computers and machines. Here we are considering a kitchen in a household.

THE AUTOMATED KITCHEN CONCEPT

Examine Fig. 2-5. The premise upon which the sketch is based is that you sit down at a serving table and speak aloud what you desire. A computer will recognize the voice (it won't serve anyone whose voice is not on record) and cause the various plates to be prepared and served at the table in the proper locations. Once the meal is concluded and you tell the computer that a clean-up is in order, the plates and all will be removed and everything will be returned to the pre-preparation state—even to the ordering of such stores and foodstuffs as are necessary to replenish those supplies used in the meal!

Look again at Fig. 2-5. There are many loops of control necessary. Here and there one will assume that there are "pick and place" type robot arms, mounted to fixed stands. These select, mix, prepare, and place foods, equipment, and utensils as necessary. As you will see there are many direct

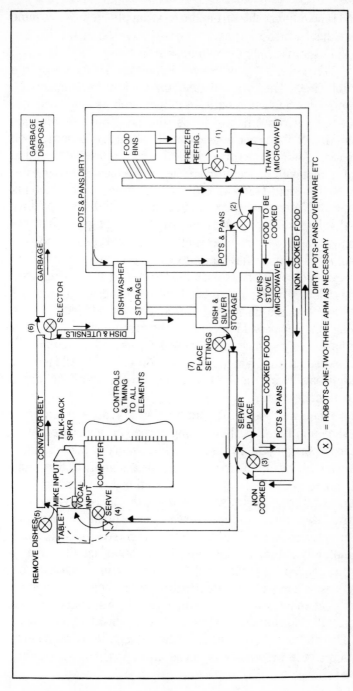

Fig. 2-5. One concept of an automated kitchen.

55

feed type units such as from the food supply bins. We assume also that somehow these bins will have been pre-loaded, either automatically or by human effort.

You need a lot of conveyor belting as shown by the double lines. Then you will need quite a few "pick and place" type robots and some quite intelligent robots which can detect the various kinds of foods, do the proper mixing and selection of plates and pots and pans and such in order to cook the food. A small microcomputer directs the whole operation and it receives its input from the people seated around the table, this input being an order for whatever kind of food is desired. Notice that the computer has a talk-back capability so that if it does not understand, or you do not give it instructions in the proper manner it can request a repeat or correction. Even if you do have an automated system it must be told every shade of every task and function that it is to perform, otherwise it will send the silver to the garbage and the garbage into the dishwasher. If you are so inclined, take a pencil and paper and improve on the system shown. It could be easier to just have pre-planned meals and assign a code number to them. When you sit down to get breakfast you simply say "Bring me A-21" and you can get your toast, eggs, bacon, coffee, etc.

When we consider the automation of a range of activities as shown in Fig. 2-5, then timing of the various processes is very important. On an assembly line where it takes a finite time for various items to reach various assembly or distributing positions, any delay in a sub-activity must mean that somehow the computer will recognize this and delay the later activities until the product has reached the position where the next activity starts. We must also, then, realize that in the completely automated kitchen (or any other activity related to this concept) that there will have to be many sensors along the way which will feed information to the central computer so that it knows what is going on every place at every instant. This will also stop the machinery in the event of a malfunction or a breakdown at some point in the system. Voice communication to the machinery is not at all impossible. It is being done all the time. I've enjoyed playing chess with a vocal computerized chess opponent and was delighted, once, when it said in its charactertistic flat tone; "I lose." I have also seen a device called a ROBOT 1, made by Heuristics, Inc. which plugged right into an Apple II computer and enabled a person to control a small battery operated car by voice commands; forward, reverse, stop, left, right and straight. The system used a radio link to the controlled vehicle and required a special program to make the computer operate correctly with the other parts of the system.

You will think of other examples of voice control and no doubt you will realize that various Universities and research organizations are working constantly nowadays to develop voice systems which respond to only a few selected individuals. This prevents troubles which could arise from having anyone being able to operate a system.The use of vocal commands to computers enables the skilled person to free their hands for more useful functions than trying to type on a keyboard. With the added capability of recognizing a persons' particular voice and taking the information that voice conveys and turning it into some action, we have the essentials of the man-machine integration in the world of robotics. Through voice communication we have found a means to affect a nearly instantaneous communication to, and with, machines; called robots, or androids, or automated kitchens!

So we consider again Fig. 2-5 and imagine how such a system would operate. Once the command for breakfast, for example, has been given to the computer, the computer then alerts the various parts of the system to be ready to comply with later commands as they are issued.

First, on the right side of the figure, is a command to the food bin that releases the proper items in proper amounts onto the conveyor belt. Commands also go to the refrigeration unit and its controlling robot (the circle with the X inside) so that as the refrigerated items are issued, the robot will place them on the belt alongside the bin order. If the food is frozen, then it must be sent to the microwave oven for thawing. This will take a little time and so a delay is now necessary. When the food has properly thawed, the robot will place it in its proper position so that it becomes a part of the order on the belt.

As the meal ingredients move along the conveyor belt they arrive at the second position. Here, those foods which must be cooked are taken from the belt, placed into the proper pot or pan which has been issued from storage. Some food items may not need cooking and so will move on along the belt to robot 3 who has the big job of combining all the prepared food and placing it in the proper dish or bowl for table use. This same multi-armed robot will place the empty pots and pans on the return belt so that they will go to the dishwasher where they will be cleaned and stored automatically. Can't you just imagine this robot #3 being a four armed, multi-digited unit, whirling and waving its arms, as it performs its assigned tasks? Of course, remember that all the while everything is going on, sensors are relaying information back to the computer about everything that is happening and when and how it is happening.

The food, in the proper dishes, the silver, etc., arrive at the table and now robot 4 begins to serve the meals, each to its proper recipient. The computer directs the order of serving by the order of voice commands, first voice is first served, etc. After a reasonable time, or when ordered, robot number 5 begins to remove the dishes in some pre-programmed order. Robot 5 puts the dishes on the converyor where robot 6 will see to it that the garbage goes one way and the dishes and silver return to the dishwasher to be washed and then proceed automatically on to the dish-silver storage shelves.

HOW TO PREVENT THINGS FROM GETTING OUT OF HAND

Prevention before cure, must be the motto concerning automated operations. Think of the modern aircraft with their many complex and intricate devices and their automated electromechanical control elements. How do they prevent mishaps? It is done through redundancy. Space robots are so designed that they have duplicate and even triplicate systems that function should the primary system fail or have a malfunction. In the coming world of robotics, redundancy and reliability will make the systems complex, but they will operate with a minimum of failures and problems. The great wizards of mathematics will manipulate their symbols and tell us exactly whether a system needs to have duplicate, triplicate or quadruplicate systems in order to achieve reasonable reliability. That means that in general the machine or system will operate correctly most of the time. It does not say that the system will ever be perfect!

In a control system the product of the reliabilities of the various sub units will make up the total reliability of the system. For example, if you have ten units, each with a reliability of 99 percent, then the total reliability to be expected will be the product of .99 taken ten times or about 88 percent. What this shows is that each subsystem must have a reliability figure on the order of 100 percent—or as near to it as possible—in order to ever get close to that 98.5 percent total system reliability. You get close to 100 percent on a subsystem by giving it duplication. Reliability engineering of robots in the forthcoming age will be of immense importance and responsibility and will attract the best engineers and scientists that our educational systems can produce.

WHAT ABOUT ROBOT PERSONALITY?

The world of robotics means an integration between people and machines and so some people have considered and made studies of what it can mean to warm blooded, emotional human beings to deal with non emotional, no-blood-at-all electromechanical devices. The studies attach importance to the types of machines which "seem to think," rather than those which are simply complicated or complex. Homo sapiens are a difficult lot, even when they are behaving correctly, so we are informed. It seems that it is much easier for these *Homo sapiens* to work with machines if the machines in some way reflect their own physical features. Thus we find that some robot manufacturers such as Quasar spend a lot of time considering how different personality traits might be displayed by their display robots.

Figure 2-6 is courtesy of Quasar, and from it you can begin to imagine the considerations that are being made to insure that these promotional display robots attract and amuse the human population instead of provoking and generating fear. It is easy to understand how a robot's appearance can range from friendly to frightening. Perhaps the day has come when the box type robot or the barrel shaped robot is *passe.*

The good engineer designs for fast operation, efficiency, low maintenance requirements, and reasonable cost. He is not concerned, really, with how the thing looks, or is packaged, unless the size and shape must conform to a particular volume for some very pertinent reason. There are packaging experts who can design appropriate housings for the electromechanical mechanisms. Perhaps they will be called in to help shape the robot machines of the future when the field of robotics demands that we have robots having personalities and bodies to match.

Consider the inanimate human models that we see in the fashionable department stores nowadays. Those models aren't cheap, and the store wouldn't buy them if there wasn't proof that they add to sales. If the same holds true for display robots or androids than we have to ask, "Just how human do we want this robot to appear?"

The real "macho" type could be as lifelike as Yul Brynner in *Westworld*, the movie in which robots run wild. The pretty type could be as attractive as those female mannikins displaying the latest fashion gowns in the finest *haute couture* salons. But, do we want them that realistic? The cost of cramming all of the machinery into that kind of package would be tremendous. Would the economic advantages of this type of robot over one that looks like C3PO, in *Star Wars*, be worth the development costs?

What about the bad psychological effects? Perhaps people feel more comfortable working with a machine that looks like a machine than they do with a machine that, due to its personalizing, seems to possess some kind of non-understandable intelligence. When a machine seems to be alive it can and probably does, generate some kind of fear in some *Homo sapiens*. That fear, of course, is usually a fear of the unknown. Since we are conditioned objects in space, and

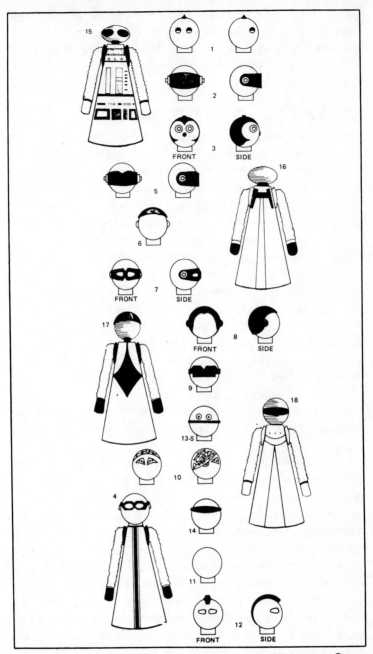

Fig. 2-6. Display robot's dress (sales promotional androids) (courtesy Quasar Ind.).

since our conditioning is a result of what we see and hear and experience, and since we see monsters on TV, etc., then perhaps the personalizing of machines may engender in the subconscious some kind of relationship which could be detrimental. Who knows?

NUCLEAR INSPECTION ROBOT CONCEPT

It is a well-known fact that one of the primary areas of robot functioning will be the ones in which mankind has trouble operating. These involve deep space, any place where there is low or high pressure or no oxygen, underwater situations, extremes of heat or cold, and environments in which there is high toxicity or possible nuclear contaminants.

A robot designed for this latter type environment has been investigated by many companies, one of whom is the Hughes Aircraft Industry with their Mobots. Another is Quasar Industries of New Jersey. It is informative to us to examine, by way of instrumentation and operation concepts, what such a nuclear inspection robot are able to do in this kind of nuclear contaminated environment, or in an environment where inspection is necessary and malfunction means contamination. With both the Hughes Mobot, which has multiple arms, video cameras, and is remotely controlled from a switchboard, and the Quasar BIOT, there is a mobility capability which permits moving the robot to an area within its mobile range.

The BIOT is a four footed, mechanical, humanoid robot. The designer says it is six feet high with a weight of 525 pounds, a width of 30 inches, and a ground clearance of some 3 inches. Its mobility is through wheels which would be self inflating and self repairing. It uses batteries for power, and the movements of its various limbs and tools and grippers is controlled by electric units with hydraulic assistance. It, too, has four arms, each with elbows and interchangeable hands. It has a means of optical reconnaisance in color, a computer control unit, and both hard-wire and radio frequency communication and control. It could use voice recognition as one means of decoding commands and instructions.

In order to have maximum operational and sensory capability the unit is envisioned as having two video color cameras in its head-unit with an auto-couple-panning capability and tilt and zoom control. Each arm bank would also be covered by a TV camera. Other sensors would be a doupler and sidelooking radar for guidance and inspection, X-ray close-scanning, Sonic close-scanning, and Geiger probes for very close monitoring. It would also have a gas analyzer as part of its sensory equipment. The tool hands could include tri-pincers, grip/lifts, adjustable torque wrenches, bi-pincer fine-tolerance grips, tool-steel pry level attachments, and a lift-hook. Then there are hot and cold hands, the hot hand capable of working in temperatures from 70 to 240 degrees and the cold one from 40 on down. Power tools attached for some kind of repair or connection include a rotor drill for ¼ to 1 inch bits, an impact wrench, screwdrivers, a vertical torque rotator, a horizontal torque rotator, a grinder, and a deep heat probe. Quasar has gone far enough in design and engineering research on this type unit to have gained a patent on the system and they believe that such units will become acceptable and used in industry in the forthcoming decade.

The chips and the CPUs of computer units are sensitive to elements of our normal environment, so one begins to wonder what kind of shielding and what type computer units will be developed to make the autonomous hostile environment robot a reality. Testing for operation, reliability and durability will be a requirement, just as we now have a requirement to produce environmental chambers in which we subject space probe equipment to various temperature radiation extremes before we launch them on their long cosmic journeys.

THE SECURITY ANDROID

We need not dwell long on the dangers from other Homo sapiens to ourselves in our modern, crowded world. Any newspaper will provide adequate information on this subject. What is important is that some kind of security is needed. This can be deterrent to all kinds of trespassers. Notice we do not say that crime can be prevented. We simply say that with adequate security the bad effects may be minimized.

Does the robot or android have a future as a security guard in such locations? Can one really be programmed to detect alien objects which may appear, and which may pose a threat to the regular and permitted inhabitants of the area? Let us see what might be required and then think of what may be possible in view of current technological developments.

If the robot or android is going to patrol an area, it must be able to do so in a random manner. It cannot patrol in a fixed operation where its appearance at any point of its path can be stated as a function of time. All an invader would have to do in this case is simply time the appearance at some point and enter during the time the robot was absent, confident that there would be a definite time interval before it reappears.

That the robot or android would follow a path of some kind in such an area is probably a requirement. It must be able to move itself to any location where its weapons or defense mechanisms can be effective against an invader. If nothing else, its alarm system must alert the proper people to come to the battle. Thus a means of causing it to follow a random walk on a prescribed path is necessary. This path, of course, must not be obvious, nor must it be easily changed or de-activated if it is an active element type of path.

One might suspect then that programming internally might be the path-following criterion. In this case there is nothing external to the robot or android which can be changed to cause it to follow an erroneous pathway, or get itself into trouble. With modern microcomputer capability it would easily be possible to select a general path and then randomize sections of it so that only the computer would know when the robot or android would reverse, backtrack, stop, wait, speed up or slow down as it follows its general trail. It should also be possible to change the movement program from day to day, or at least from week to week so that one could not determine a fixed pattern, no matter what it might be, by watching and counting and timing the machine's movements. Also, by the very definition of randomization, the machine's movements along a general path could not be predicted.

The area would also be safeguarded by some kinds of fixed sensors, such as terrestrial vibration units, which are now available in such sensitivities that they can detect a fox walking on the earth at distances of over 10 feet. Infrared beams of light, invisible to the human eye, can also be used to set up a penetration barrier, and all such devices can be easily arranged so they would signal the robot or android the moment they detect anything foreign to their habitat. The robot then receives an over-ride signal which causes it to speed back to the appropriate location immediately.

It would seem that quick penetration of such barriers might be accomplished. It would be getting out that might prove difficult. Once the sensors have determined that an invasion has taken place, a sufficient number of security robots could immediately move into a tight and overlapping coverage position so that escape from the inclosure would be impossible. Disabling devices could then be used. This could go as far as the aiming and firing of a gun-like weapon. Laser light beams can cause a small weapon to be aimed and fired, just as they cause the big guns on tanks and aircraft to aim and fire. Laser light, blinding in intensity might be used by first causing the intruder to look at the robot. Sound waves, the emission of gas, and other methods might also be used.

How would one pass the sentry if one had permission to enter the area? Perhaps with an automatic device such as is used to open and close garage doors. Pressing the button would tell all devices that a friendly alien was coming inside the guarded area. Thus they would be permitted to pass.The code used in this case would have to be a carefully guarded secret, and might be the only weak link in the chain of security devices. It is a challenge to come up with a security system so that some can pass and all others will be rejected, using some means of distinction which cannot be compromised.

We have already mentioned one device which might be used in the sensory role; the infrared light beam and receiver. Another unit which might be adapted to this role is the microwave door opener, manufactured by Solfan of Mountain View, California. The unit projects a beam as shown in Fig. 2-7 from its microwave motion detector. When it sees anything moving in this beam pattern it activates a relay which, in this case, causes the door to open, but which in our application

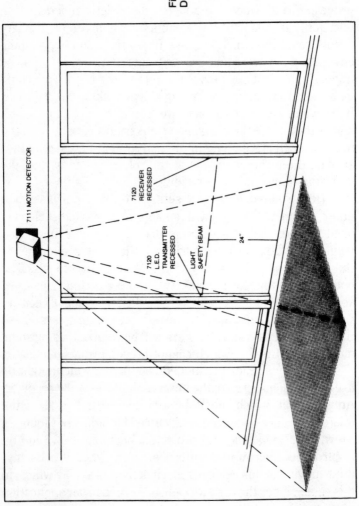

Fig. 2-7. Solfan Microwave Motion Detector (courtesy Solfan).

7111 MOTION DETECTOR

7120 RECEIVER RECESSED

7120 L.E.D. TRANSMITTER RECESSED

LIGHT SAFETY BEAM

24"

might cause an alarm to be sounded, or a robot to be summoned. Also indicated in the diagram is a light beam unit across the doorway which prevents the door from closing in case a person stops in the doorway. Remember that the motion detector operates on a doppler frequency shift caused by a moving object. Thus, when the object stops moving, there is no signal to the activating relay. The microwave detector in this system operates on one of four channels around the 10.525 Gigahertz region. The relay is activated by a doppler frequency shift when a moving object penetrates its beam. The beam pattern can be adjusted for width, length, and depth. It is sensitive to movement from one inch per second to as fast as sixty inches per second. It operates on 12 to 24 volts ac or dc. Its detection sensitivity can be adjusted using a potentiometer control. Some of the antenna coverage patterns are illustrated in Fig. 2-8. If you are a robot designer you will no doubt think of many ways in which this little unit might be used.

I offer one suggestion, illustrated in Fig. 2-9. Here, four radar units are used to give the robot some self defense features. No one could approach this robot without activating its motion sensors and causing a triggering of its defense mechanism features. One must realize that if a robot is used in a security role, it must be able to defend itself, or at least alert its master to the fact that it is being attacked.

PATH CONTROL CONSIDERATIONS

While it is true that some robots have been built that will follow a prescribed path they do not all use the same type of path control. Some use a computerized path specification wherein the path is put into the computer's memory and then the computer activates the steering section in accord with these pre-determined, selected commands. The other method is to use a type of cam. You can well imagine how such a cam following system might be built. A rotating mechanism of some type must turn a shaped cam so that it will make contact with microswitches that in turn steer the robot. This cam arrangement is very simple and reliable. It is not subject

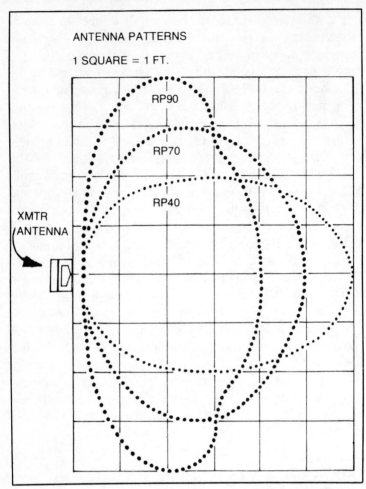

ANTENNA PATTERNS

1 SQUARE = 1 FT.

RP90

RP70

RP40

XMTR
ANTENNA

Fig. 2-8. Solfan Microwave Motion Detector Antenna Patterns (courtesy Solfan).

to interference and may be changed quite readily by changing cams.

If we expand on the cam concept we can imagine a machine which will shape a cam or cams as we push it around a specified path. Then we simply have to put that cam into the robot's mechanism and it will follow that path which we have laid out, assuming, of course, that nothing happens along the path to cause the timing of the cam steering to be incorrect relative to the robot's present position.

68

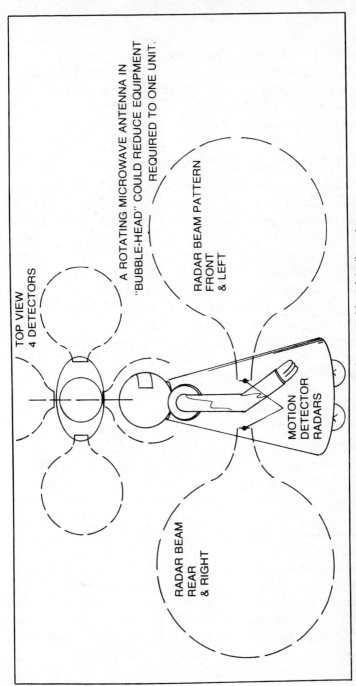

TOP VIEW
4 DETECTORS

A ROTATING MICROWAVE ANTENNA IN "BUBBLE-HEAD" COULD REDUCE EQUIPMENT REQUIRED TO ONE UNIT.

RADAR BEAM PATTERN FRONT & LEFT

MOTION DETECTOR RADARS

RADAR BEAM REAR & RIGHT

Fig. 2-9. A robot might use radar motion detecting units for self defense, or object detection, etc.

69

One can also imagine certain subtle variations in this cam steering, motor-controlling concept. One might use magnetic tapes generating tones of various types for various actions. It would seem that with such programmed steering and speed control, a house-movement robot or a lawn and yard and garden movement robot can easily become an autonomous reality!

A TIMED PROGRAM FOR PATH CONTROL

When we consider a robot following a particular path, be it in or out of the house, we must remember that a robot will be asking for some timed directions. A timed direction is simply a signal which says the robot should go for a *specified* period of time. If the speed of movement of the machine is known, for example assume one foot per second, then you, as Robot Master, know that if you let that signal exist for five seconds the robot should move five feet. I say should because there might be some impediment to the machine's motion. We begin to see that programming a robot for motion along some kind of path, specified only by the internal timing and speed of the mechanism, is not feasible.

COMPUTERIZED PATH CONTROL

How do we achieve computerized control of the robot for the same path? Well, we first plan to have some kind of feedback to the computer so that it can tell when the machine it is controlling has done what it is supposed to do. For example, assume that we have some kind of sensor which will send one pulse per wheel revolution to the computer. One revolution will mean that the wheel has traversed a distance of $2\pi r$ inches where r is radius of the wheel in inches. When the computer sends a signal for the robot to move forward, it will be getting back a distance measurement pulse from the wheel monitoring sensor. The computer can count these pulses to determine distance. You might even have a pulse generated for each quarter turn of the wheels. The computer memory is used when you load it with software instructions that tell it the number of pulses for each segment of the path.

The computer will be constantly checking against this number with the direct input of pulses from the wheel sensors.

There still could be some difficulties with this type of control system. The wheel could slip, indicating distance traveled, even though the robot didn't move. And what about when the robot is turning? If you assume one wheel is a pivot for the turning motion, then you can quickly see that the wheel might rotate even though it is going nowhere. You have to think about the turning problem and plan for it. The slipping problem cannot be solved using wheel sensors.

MORE ON THE MICROWAVE MOTION DETECTOR-SENSOR

Consider the use of a device such as the Microwave Motion Detector made by Solfan. AHMAD, a robot made by computer technician Donald Dixon of San Jose, California, is purported to be able to monitor a children's room. If they try to leave that room, the robot tells them not to, and if the robot encounters a prowler it will tell the prowler to go back. This robot is also programmed to accept and deliver messages.

Our robot, when equipped with the radar motion sensor-detector, can respond in a variety of ways to anything moving in its field of view. Since the radar unit is normally designed to activate a relay it would require a little change to activate a verbal questioning unit, cause a light to come on, or cause a loud barking to sound (from a tape).

If someone responds and the voice is not recognized by the robot's speech identification circuits, a siren might sound or lights might flash with a horrifying intensity.

There are good speech recognition and speech synthesis circuits currently on the market, and more are being developed and these will certainly be used in robotic applications. The limit of what kind of talk might be possible with speech-recognition and speech-generating circuits is limited only by a person's imagination.

Use, Operation, and Construction of a Radio-Controlled Robot

3

He is called GARCAN, and he is manufactured and displayed by Jerry Rebman Electronics Co. They can supply all parts for this type robot. He is a toy-like being that is capable of wheeling around at parties and on display floors, under the able and capable radio control guidance of his master, Jerry Rebman, and he can converse with attendees either in groups or single units. He can entertain with songs of any type, or other 8-track recorded sound, through his concealed, high fidelity speaker system. His purpose in life is to entertain and to do this, like his human actor counterparts, he can change his costume, and thus his identity, as simply as donning a new outer shell covering.

GARCAN is a rugged unit and can withstand the pressures of children climbing up on his "garbage can" working costume without sustaining damage. He has provided amusement and entertainment for the most elaborate of high society affairs, as well as for those youngsters who are confined in hospitals for all the well known, but little understood, reasons. His manner of speaking, at the moment, is through a intercom unit which permits his master to listen to and talk back to anyone who says anything to GARCAN when he is "on the job," except, of course, during those moments when GARCAN's internal tape player has been activated.

My first encounter with this delightful unit is illustrated in Fig. 3-1. I had been invited to the home of Jerry Rebman and after a few moments he vanished outside this lovely family room. A moment later, to the tune of purring electric motors, GARCAN appeared in the doorway and with uncanny accuracy and agility entered the room, flashed his "head" lights at me, decided I was friendly and came closer (Fig. 3-2). There he stopped and silently examined me and all the while I watched the maze of lights flash like small bits of lightning under his plastic, hemispheric, domed head-space. I noted his rugged construction and how well the inverted 30 gallon garbage can made a suitable body when mounted on the circular, one inch thick wood or composition board base platform. I could just barely see the front idler wheel which gave the base platform good stability. I assumed, and correctly, that there would be two other wheels located further to the rear of the platform and that these would not only provide mobility, but also steering. I also noted how easily GARCAN moved across the bare floor and the thick carpet.

There are countless types of display robots being used now at expositions, fairs, on TV, in movies, at parties and such, and it is safe to say that 98% of these are radio controlled types. They may be more elaborate than GARCAN, who is a paraplegic (no arms at present) because the addition of arms means much more complexity. Others may have arms and wrists and hands, but these are all operated by control signals sent from behind one-way windows in exposition booths. Display robots of this type may also have a multitude of self contained sub-programs which can be started or stopped by radio signals, and these may range from a pre-programmed path movement, to animated arm and finger movements, to speech delivery, to music, to display of TV pictures on a special built-in screen. We have found that the big companies such as GE and Hughes and others have used such display robots to present their messages or logos to the public at various functions.

Display robots are a reasonably big business and will be a larger business activity as time progresses. They cost from

Fig. 3-1. "Hi There! I am GARCAN."

Fig. 3-2. "Well, look me over!"

4 to 10,000 dollars each and they rent, with a Robot Master, from $50.00 an hour upward after expenses are paid. Some types are "quickly-put-together" units for profit, and some

are well built scientific experiments. In the display robot role, these units have but one purpose for their existence; to entertain, amaze, and delight those who see them. One way in which this amazement is encouraged is to add flashing lights or other eye-catching devices which are clearly visible, under protective covers, and which seem, or even may really be, a part of the robot's internal control system.

We examine GARCAN closely because he gives us a good insight as to what makes one robot "tick." He is a well engineered model, and a person might expand his basic system to include a multitude of sensors, computerized control, TV eyes and such, if one is inclined toward experimentation or design in this direction.

EXAMINING THE INSIDE OF GARCAN™

Now we examine the inside of GARCAN. It is easy to loosen three bolts which hold the inverted garbage can in place on the circular base structure. There is a multitude of wires running from the base electronics to switches and electronics inside the bubble-head. One has to be careful to lay the garbage can down right next to the base as shown in Fig. 3-3. Another view of the two parts can be seen in Fig. 3-4, and a good view of the internal wiring is shown in Fig. 3-5. Notice that there is lots of space should one desire to add arms and the arm mechanisms. No attempt had been made to cable the wiring and there are no special wires except one. A wire from the radio receiver, located in the bubble, which comes down to the motor control unit is a coaxial type. Coax prevents stray signal pickup which could influence the control, steering, or drive signals to the base mounted elements. In Fig. 3-6 Mr. Rebman points out a particular circuit holding a standard radio control transmitter such as made by Heath Co., ACE Electronics, Kraft, etc. The second robot is wearing an armadillo suit covering. The electronics for the two robots are identical, but different RC frequencies are used for control so both may be operated simultaneously.

There are many other covers which can be used with the base platform. One is a snowman, another is a Christmas tree.

Fig. 3-3. Top of GARCAN removed from the base.

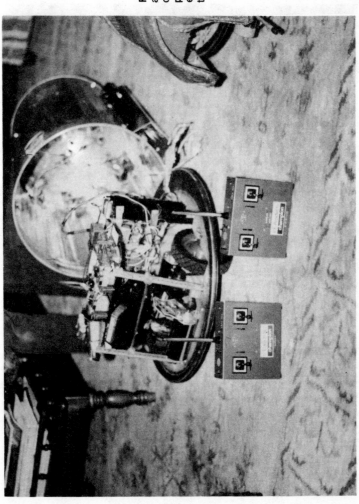

Fig. 3-4. View of base equipment and radio transmitters used to control GARCAN. Transmitters are parts of commercial radio control equipment, for model airplane control.

79

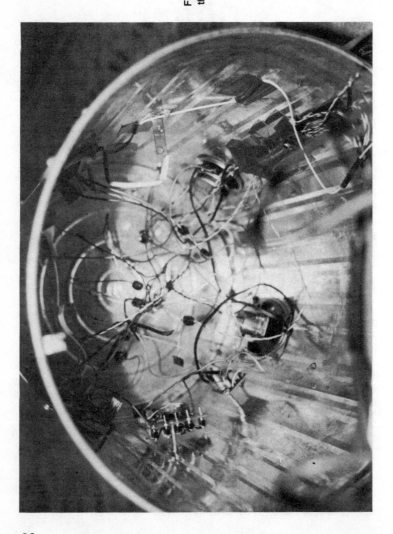

Fig. 3-5. Sparce wiring is inside the garbage can cover.

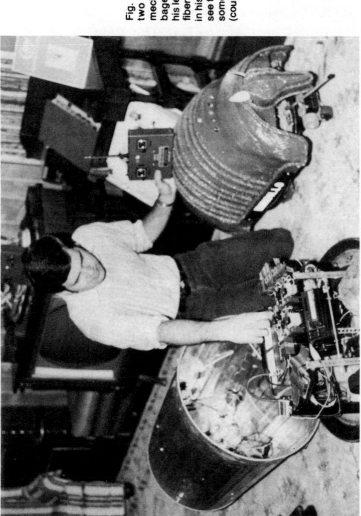

Fig. 3-6. Jerry Rebman shows the two "tops" for the robot base mechanism. On his right is the garbage can cover normally used, on his left is an armadillo cover made of fiber glass. The control transmitter is in his left hand. On the base you can see the 8 track tape used to produce some music or other vocal sounds (courtesy Jerry Rebman).

The armadillo covering is made of fiberglass, and on this robot there is a forward jutting arm which is used to prevent the robot from damaging its head if it happens to run into something. On the reverse end of the covering is a tail. This has no electronic purpose or defensive purpose but is for effect only. Because he is such a pet, the smaller armadillo has been given the special name *Minidillo*, and he is shown in closer view in Fig. 3-7.

INSIDE THE ROBOT BASE STRUCTURE

We look at a view of the base structure in Fig. 3-8 and we can see how one drive motor and two steering wheels have been mounted. The wheels project through the base. We can see that the one on our right is connected to a electric bicycle motor (¼ horsepower) by means of a chain and some sprocket gears. Each wheel you see here has its own motor.

With the swiveled front wheel, the three wheels will form a triangle with a wide enough span so that very good stability is achieved for the vertical structure. Also, not visible in this photograph, but present, is a deep charge 12 volt automobile type battery which is located just behind the front wheel and just to the rear of the tape cartridge shown. Jerry informs us that it is very important to get a deep charge type battery as the motors draw some 30 amperes when stalled, and about 5 to 6 amperes when running, and so for operation over any length of time a really good battery is needed to supply motive power. This same battery, with voltage reduced and regulated to a value of about 4.8 volts, is used to power the radio receiver, and the radio control servos. Switches control the larger power flow for the tape deck and lights and such.

Since motive power is such an important first feature of any robot of this type, we get even closer through Fig. 3-9 to the motors and drive mechanism. Each rear wheel has an identical chain drive and an identical electric motor of the type shown. The speed of each motor is controlled independently by radio command through the radio control transmitter held by the Robot Master.

Fig. 3-7. A smaller version of the armadillo is the Minidillo. Note switches, and lights on its shoulder to indicate when base is powered.

83

Fig. 3-8. A close-up view of the base platform mechanisms. Note the drive motors and drive wheels just below the 8 track tape unit. Two drive motors are used similar to the one in the foreground.

84

Fig. 3-9. A still closer view of the mounting method for the electronics and mechanical units. Realize that each of the driven wheels has a short axle so a good mounting is necessary.

I wondered if it was a problem to synchronize the motors to get forward travel. This, said Jerry, is not a problem. Since the motors are nearly the same, and since he has independent control over each motor's speed, he can easily—with practice—adjust the RC sticks so that the robot moves forward or reverses in a straight line. To make the robot turn, you simply advance the speed of one motor while you reduce the speed of the other motor. A tight turn results. It is also interesting that a double gear-reduction via the chain drive sprockets, is used to get the speed of motion down to about a fast walk. Using a gear reduction you increase torque to the wheels and thus the unit is able to operate over thick carpets, grass, normal unevenness in concrete and such. A single wheel for both motive power and steering of such a base platform is not really practical except on smooth surfaces. The arrangement shown here has turned out best and will prevent tipping of the robot even on a steep grade.

A good side view of how the tape player is mounted, at an angle above the motors, is shown in Fig. 3-10. You can see that when a given tape is placed in the unit it cannot be changed without removing the body from the robot. This presents no problem, really, as you know ahead of time what kind of tape you want inserted, and the tape is long enough so it doesn't get boring during a demonstration. Another advantage is that *you* can record your own jokes, stories, or whatever. With the 8-track unit, there is radio-control switching of tracks. You know what's available on all of your tape, and you simply command whatever track is appropriate at that moment. The commands for this function are sent to the regular radio-control servos which are mounted so they open and close the microswitches which operate the tape player, a very reliable and positive method of control.

We can examine the gearing and chain drive system still closer in Fig. 3-11. Notice that the driving wheels do not turn to accomplish steering. The wheels are fixed in the plane shown by means of rigidly mounted axles. To turn the base, one of these wheels is made to go faster or slower than the other. To the right of the nearest wheel is a black looking

Fig. 3-10. Installation of the tape player at an angle. Heavy duty battery in upper left, some mounting brackets for electronics boards, details of the left drive wheel show bolts used to fasten gear to the wheel.

Fig. 3-11. A close view of the chain drive system for the wheels. Note that *two* chains are used on each wheel. You can see the two motors in this photo if you look carefully. One is just behind the chain-driven gear in the center.

vertical stack. This is one of two heat exchanger mountings for the powering transistors for the motors. These transistors get really hot and must be mounted on a heat dispersing base. The motors themselves also get very warm to the touch, but they operate within the manufacturers specified temperature limits. The wheels are obtained from a hardware store and the gear is bolted to them. Bolt shafts can be seen on the nearest gear, while in the background you can see the relative size of this gear. The chains are smaller than the regular lawn mower type, but very strong.

We have mentioned the use of the regular model airplane servos as mechanisms to control function switching. In Fig. 3-12 we get a better idea of how this was done. By close examination you can find the little servos with their plastic arms which, when moved by radio command, open or close the microswitch levers on each side of them. These servos and switches are bolted down rigidly to a base plate of plastic and this, in turn is fastened down to the supporting structure above the drive motors. The use of microswitches is important. Since they snap into contact or out of contact, they make a positive connection quickly and thus may prevent burning or pitting of points. Also, since they have a good rating in conductivity, the points are large enough to pass all the current required for the auxiliary items easily and without loss.

Because the driving section of such a robot is usually the biggest problem to those who would fabricate a robot, we take another look at the drive motors and their mounting and the chain link to the wheels in Fig. 3-13.

In this figure you can see, with a close observation, that each end of the drive motor has been equipped with an L-shaped mounting bracket. This, in turn is bolted down to the baseplate. The use of such a bracket, which is bolted to the motor, prevents the motor housing from rotating as the armature turns. Sometimes you can use a cylinder case on a motor and a strap around it to hold it down but if the strap loosens, you have a motor housing rotating, or trying to, and this causes problems. It is better to bolt a bracket to the motor housing at each end of the motor and then bolt that

Fig. 3-12. The radio-control servos are mounted so when the arm moves it closes or opens, the micro switches.

ONE SERVO

2 MICRO SWITCHES

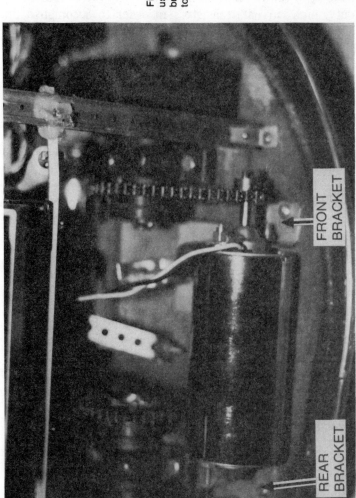

Fig. 3-13. Details of motor mounting using brackets fore and aft. These brackets are hand made and bolted to the motor case.

FRONT BRACKET

REAR BRACKET

down to the baseplate as shown. Observe, also, that this is but one of the two drive motors required for steering and motive power of the GARCAN robot.

I have said that the drive transistors used with this motor controller electronics system need to be able to pass high currents which exist when the motor is starting and when various speed changes occur. Going from a stopped position to a full speed (approximately one foot per second) requires a surge of about 30 amperes as opposed to the 6 amperes needed to run the motors. Each motor requires four heavy duty, high current transistors in order to accomplish the switching required to permit variable speed control and reverse motor rotation. These transistors take the place of relays and are more effective than relays in this application. Since the transistors must be mounted on a heat absorbing and distributing plate, they are mounted separately from the printed circuit board used for the balance of the motor control electronics. We can see four such transistors properly mounted on a vertical metal plate with fins for heat dispersion in Fig. 3-14. A terminal strip is bolted to the left side of the plate to make connections positive and easy. These transistors get very hot.

The balance of the motor control electronics can be seen in Fig. 3-15, in the center of the baseplate section. The heavy duty 12-volt battery is normally located in the open space toward us and away from the motor. The motor is the second of the two required, and above it you can see the two-section printed circuit board which contains all the necessary electronics to use the signals sent from the radio control transmitter to its receiver—a series of pulses which change their position with respect to one another and a reference in accord with the stick movement on the transmitter. These circuit boards convert these pulse positions into pulses of electric current and feed these to the motors at a much higher current level than the receiver is capable of delivering. The number of these pulses per second determine the motor speed, and the polarity governs the direction of motor rotation. Since two drive-steering motors are required on this robot, two com-

Fig. 3-14. High current motor control transistors on a heat sink base which also serves as a support structure.

plete motor-control circuits are required, using 2 radio control channels.

Notice that the mounting plates for the transistors are located at each end of the photograph. They are also used as a support for the second story deck on which the small radio-control servos and microswitches are mounted, and to which the tape recorder is fastened.

Fig. 3-15. The motor control electronics boards may be purchased (Rebman Electronics) or built. Some types of bicycle electric motors which might be suitable also can be obtained from Rebman Electronics or General Engines 5446 Mantua Blvd, Sewell, New Jersey 08080.

Before examining the circuit concept, we take a look at Fig. 3-16 for another variation in this type of robot. This is a Tall Boy unit with a high square body mounted to the base-plate covering, and with its head bubble located at the top. It stands some six feet tall, is painted black and when it comes into a room it does create some excitement with its flashing lights and strange noises!

Fig. 3-16. Jerry Rebman's 6 foot robot (courtesy Jerry Rebman).

We have stated that these robots are paraplegic—no arms. If one decided to add arms, it would require additional motor control circuits and motors. One motor and circuit board would be needed for each arm as a minimum. With this much additional equipment, the robot could be made to raise and lower his arm—or with two such circuits, both arms either singly or simultaneously. No elbows or wrists or grippers would necessarily be used. Two more radio channels than the minimum of three (for motor control right and motor control left and tape unit) would be needed. But one might use an 8-channel RC system (they are available) and cause the closing of more microswitches to open and close pincers etc., if one wanted to experiment in this direction. Also, don't forget that a person might use an 8-channel RC system for many functions with *that* RC system operating on one frequency and then *incorporate a second RC system with another 8 channels* operating on a second frequency! Since it would operate on a different frequency no interference would be experienced between the two. As we indicated earlier, the limit as to what you can do with your robot will simply be limited by time, effort, and money!

OPERATION OF GARCAN ELECTRONICS

Figure 3-17 shows how a computer may be interfaced to drive a small radio-control type servomechanism. The importance of this circuit is that with it, you can interface a computer to control your robot in either an autonomous or a remote state simply by adding additional sections to the GARCAN motor control circuit. Notice that the output of the ICI 556A is driving a standard model airplane 3 wire servo. It provides as output, the type of input necessary for the motor control circuit of one motor of the robot. You would have to experiment with the addition of a second similar circuit, connected to the computer output, or to the MC 14081-8 chip to make a second channel to drive the second drive-steering motor circuit of the robot.

If you are ever to have an autonomous type robot, it will have to have some kind of computer within itself. A domestic

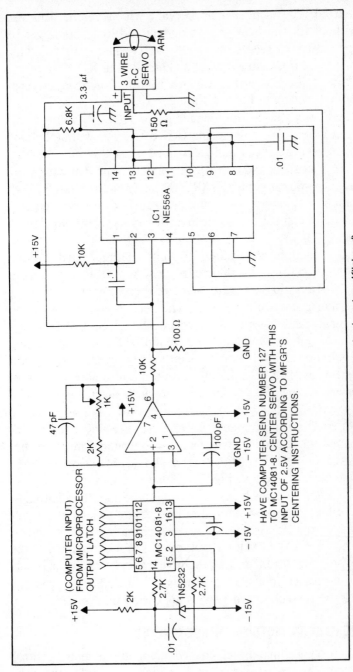

Fig. 3-17. Driving a model airplane-type servo from a microcomputer (courtesy Kilobaud).

97

android type of household servant will never be achieved without it! The best way to understand how this might be applied to the GARCAN robot, is to examine the following circuits. First take a look at a sketch of the RC transmitter. From it learn how the various robotic functions are controlled. Examine Figs. 3-18 and 3-19.

Notice that even the voice is controlled by a stick position. When you are steering GARCAN, you will be holding this type transmitter and you will be using the thumb of each hand to position the two sticks. Jerry tells us that with some practice you can control GARCAN easily and accurately. Both thumbs would be working simultaneously as you can imagine. All radio control bugs will love this type robot! Now study Fig. 3-20.

It might be of help for us to point out that if you do not use the radio control link, then there must be two signals from the digital-to-analog converter. One of these signals must go to the motor controlling circuit. The second signal must go to several standard type RC servo units which control the off-on switching for other functions. A motor for this application should be relatively small, be either permanent magnet or a field wound, 12 volt, direct current motor of about ¼ horsepower. If the robot you intend to power is much heavier than, say, 100 lbs, then you might think toward a 1/3-horsepower motor. Each motor must be capable of being operated at normal speed and load with an input of around 6 amperes, and its stall current or starting current should not exceed the 25 to 30 ampere range specified for these two motors.

Since we have been discussing the computer addition to the robot control system, it is interesting to look at a small microcomputer system and its disc operating system shown in Fig. 3-21. You can imagine typing in your commands on the keyboard and transmitting them via radio link to your robot, and if you let your imagination work a little harder, you can imagine the robot's responses appearing on the CRT!

THE GARCAN MOTOR-CONTROL CIRCUIT

Jerry Rebman Electronics developed this motor control circuit using as a basis, a circuit which appeared in the Feb-

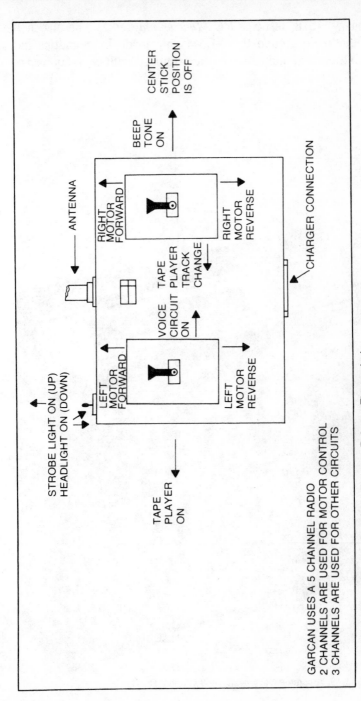

Fig. 3-18. GARCAN transmitter functions (courtesy Rebman Electronics).

99

ruary, 1978, issue of *RC Modeler magazine*. The circuit is developed around the NE544 chip made by Signetics Inc. Extensive modification to the original circuit was required by

Fig. 3-19. A commercially available 8 channel RC xmtr.

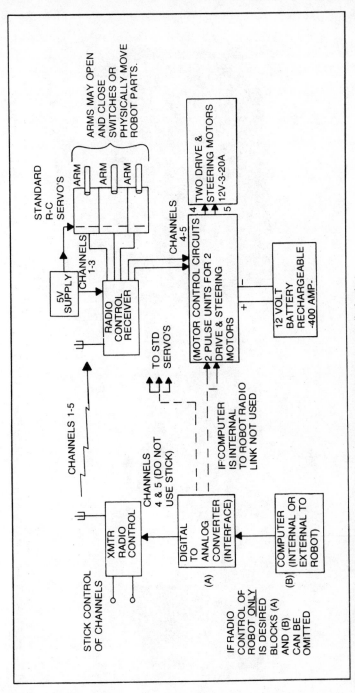

Fig. 3-20 A General block diagram of an RC directed and computer controlled robot.

101

Fig. 3-21. A Polymorphic computer system (courtesy Polymorphic).

Rebman Electronics, and it is with their permission that we examine it here. In Figs. 3-22 and 3-23 we can see the circuit layout.

From the left side of the diagram, we find the input which has been prepared for positive pulses. Some RC transmitters and receivers use a negative pulse arrangement. There are circuits which convert those negative pulses to positive pulses and ACE Radio Control is one place where such conversion circuits might be obtained.

Next we find the ground, or common connection and at the top the 5-volt positive input supply line. Notice that the ground connection is common to both the 5-volt supply and 12-volt supply. Both supplies *must* have a common ground.

The input pulses, varying in spatial position, or time position, in accordance with desired commands, are input to terminal 4 of the chip. Thence the chip's connections are made to various resistor-capacitor arrangements among which will be seen the trim pot and the range pot, two important adjustments. The trim pot is used to set the motor speed to zero when the transmitter control stick is at neutral. Refer back to Fig. 3-18 and see how the motor changes its direction of rotation as the stick is moved up, or moved down. When both sticks are at the center or neutral position, neither motor runs and so the robot stands still. The range pot governs how quickly current is applied to the motor. This, then, in turn, governs how quickly the robot will start to move when you move the stick from the neutral position. You simply have to adjust this potentiometer to your own requirement. You don't want the robot jumping forward, and you don't want to have to wait a long time for it to start moving. So you adjust the range pot for a condition suitable to you.

The switching of large currents to the motor is accomplished by the heavy duty transistors to the right in the diagram. Two 90548 types are used as drivers, and the others control the flow of current to the motor windings. These transistors take the place of relays—if you want to think of them that way, but they do a little more than a relay can do. They can control the amount of current passed, as well as the direction of flow of that current. A relay cannot do this. The

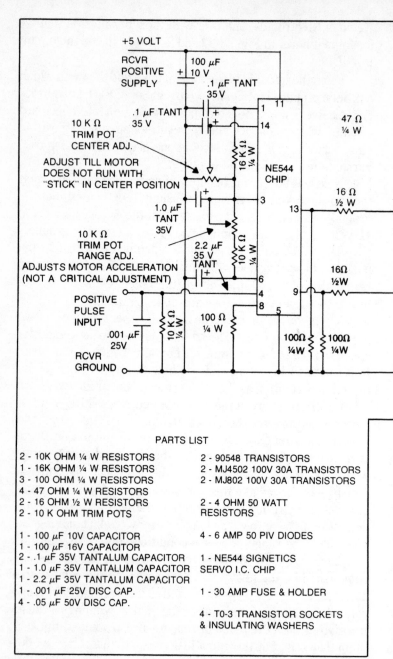

Fig. 3-22. The Motor control circuit for GARCAN. Two such circuits are required, one for each drive-steering motor (courtesy Rebman Electronics).

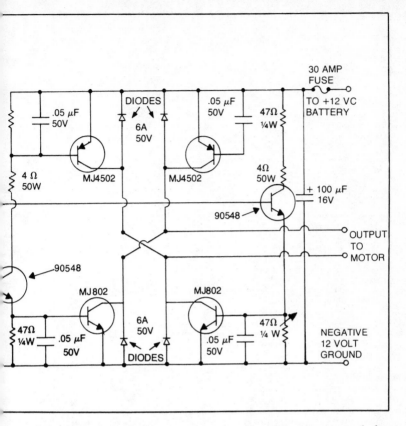

relay can turn the current on or off and it can control the direction of current flow, when properly wired, but it cannot control how much current passes its contacts. So, we have direction of rotation and speed of rotation controlled by the heavy duty transistors shown.

Finally, on the construction side of our discussion, we note that in the following chapters we will supply information about robot eyes, brains, and mechanizations, which may be of help to you in expanding on this basic unit, should you desire to do so. While it will be fun to operate the unit as is, still it is possible that you might sit back and imagine your robot doing useful tasks and jobs for you. To that end it might need arms and all the joints and appendages which will make it useful in a working capacity. Computer manufacturers have developed systems which can be interrogated and operated

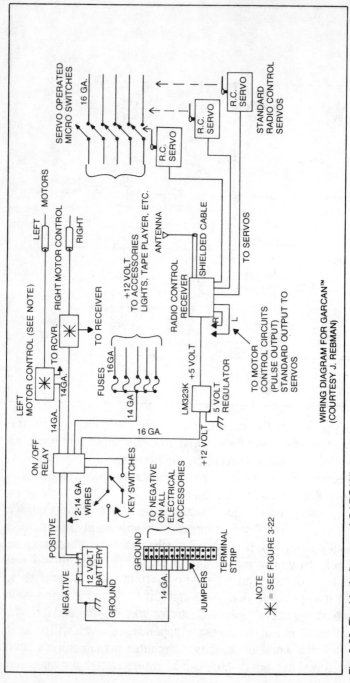

Fig. 3-23. The block diagram of GARCAN's control system.

106

via telephone from anywhere. It is not beyond belief that one could have a home sentry robot in the house when one leaves home for a trip that would report by phone!

Robot Control Programming

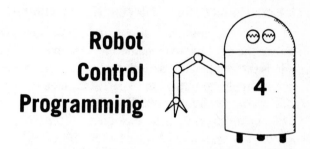

4

In the previous chapter we have discovered how a computer might be integrated into an automated control system (a robot) and we assume that the computer will take over and direct the machine. This is what we want it to do. We also have, deep in the recesses of our minds, the idea that the computer will have been programmed *by us*, so that the machine will do the things we want it to do. This may not be a revolutionary idea, but it will be our working principle concerning the incorporation of the computer into the system.

In this chapter, then, we want to try to examine some of the ways in which we program such a computer. What types of languages are used for control? How do they differ from other computer languages which we might be familiar with? We shall not restrict our examination merely to the hobby type machines but will expand our knowledge by investigating methods used in the commercial world of robotics. In a majority of cases we will find that what applies to one situation will apply to another.

As we well realize, one of the closely guarded secrets of the commercial robotics companies may well be how to communicate with a particular machine. There is a good reason for this. The simplicity or the complexity of the language governs the simplicity or complexity of the machine itself.

The software used with a commercial robot may be a vital key into its most guarded operational secrets. New patents are filed constantly. And just as there are new machines arising, there are also new methods of controlling these machines. For a variety of reasons, some manufacturers are reluctant to reveal these methods.

A PRELIMINARY LOOK AT MACHINE COMMUNICATION

Since our communications to a robotic machine involves telling the machine to do something, we must first have in our minds the sequence of actions that we want performed. Let us use a simple definition to expand that idea of action. Let us call whatever the machine does an *event*. If it makes a light flash, or it causes a visual blob on a TV screen, or it makes a motor turn ⅛ of a radian, it is an event. It is the outcome of having the machine perform an instruction, which we will convey to it in some kind of language that it can understand. So, one of the first premises that we must make is that we can take something we understand and this will—through a conversion device of an electronic nature—become an electronically coded signal which the machine can understand. When we can do this, we can communicate with the machine.

Let us think of a simple example for a computer-machine system. Suppose we choose a simple lettering arrangement such as GF to mean to us that the machine must engage such mechanisms as necessary to start the machine moving in a forward direction. We must be clear in what we mean by forward in this situation. If we have a machine which is mobile (a robot such as GARCAN) this letter code would mean that the drive wheels start up and both rotate at the same speed in the same direction moving the robot *in the direction it is facing*. We must have agreed with ourself that this is how we will define forward because there is no other reference used in this robot system except its own body geometry. Notice we have assumed that the robot can "face" in a direction! We derive this from the human body which has a "facing" direction and a "rear" direction!

The letter symbols GF mean absolutely nothing to the machine until they are translated into its language. The machine uses electricity as a working element, and so GF must convert to electricity in some unique and distinct manner so it cannot be confuse with any other electrical signal(s) which might be present in the system. The computer can take the GF input from two keys on its keypad and from its memory section produce, as a function of time, a very specific series of pulses when each key is pressed, which, in turn, represent a code inside the machine, which can cause things to happen. As you know a pulse, is known as a "one", and the non-existence of a pulse, within a specified time interval, is known as a "zero". Each stream of pulses for each key is unique. In computer language a machine word has been formed.

Let's make the case easy by assuming that we get a code like 1111 when we depress the key with the letter G. On the way to the big event, the four ones go to certain other circuits. The four ones energize—in this example, four gate circuits which will energize some solid-state or physical relay circuits, which could cause events to take place mechanically if and when another code signal of the machine type energizes the second input to the gate. So we now depress key F and release another string of coded signals. This time we have designed the memory bank to release a signal which looks like 0001 three zeros, or no pulses, and then a pulse, all in the proper time relationship. These signals also go to those four gate circuits. You've already guessed what happens: Three of the gates having a one from the first letter now get a zero from the second letter and so these gates do nothing! But that fourth gate gets a one from the first letter and a one from the second letter and so it activates its relay and that relay then applies electrical power to the drive motor system of the drive motors in such a way that the machine moves forward. Now we can begin to imagine a program to cause many events to take place. Such a program would simply be a series of two letter commands, taken in proper sequence.

When we think of a program through a computer we must also consider a time factor. In the previous example, how long do we want the machine to go forward? If it is for a specific length of time, such as, for example, the time it takes the arm of an industrial robot to swing from an end position to the start position, then we must either have a clock timing the arm movement or a sensor at the end position which, when activated will send a signal the machine can understand, back to its computing section to tell it to stop sending forward signals out of its memory.

So, what's the problem? The problem is to have a computer that can convert the type commands we wish to issue into suitable control signals for the mechanism to produce events. Our commands may be of the immediate input form, such as when we type them on the keyboard, or they might be a stored typed on disc or cassette if we decide we don't want to sit there and command that unit all day! If the machine is to perform many complex movements or operations or events, then we will need subroutines performed in the proper order and sequence so that no one—not even ourselves—can get all mixed up and forget or omit some command which is necessary to the whole operation. Very, Very, careful and accurate programming, complete, and without bugs is necessary for control of machines! Figure 4-1 illustrates this discussion.

WHAT ABOUT THE ROBOT THAT LEARNS?

We have mentioned a type of robot machine which is taught what it must do by having it go through the series of required movements and actions under the direct control of a human teacher. One such robot is the Cincinnati Milacron T^3 Industrial Robot, shown in Fig. 4-2 doing a commercial welding job. The robot is an articulated arm, and here it works with a robotic companion, to the left in the photo, which is a small squatty unit which turns the work to various positions as required. We know, just by looking at the photo, and with our limited knowledge of welding, that the arm must be capable of moving that welding wire and torch to just exactly the right place for a spot, or along just exactly the correct line for a seam. It must deposit just the right amount of weld material, with the proper heat, for a perfect joint. It does this under the guidance of a computer.

Before moving further into the operation, let us examine the physical aspects of this arm in more detail. This is benefi-

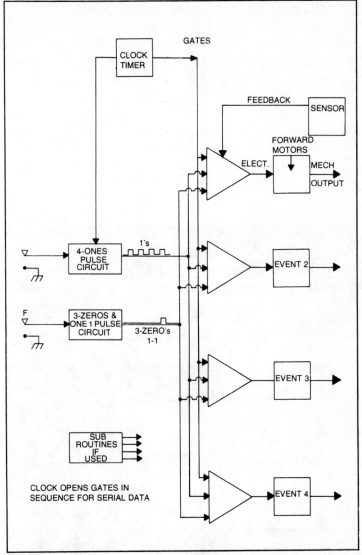

Fig. 4-1. An elementary concept of a computerized control system.

cial because this type robot is in great demand in industry, and as the demand for goods built to rigid specifications increases, more and more of these type robots will be found in the plants and manufacturing centers. They will do much more than just welding operations or paint spraying. These

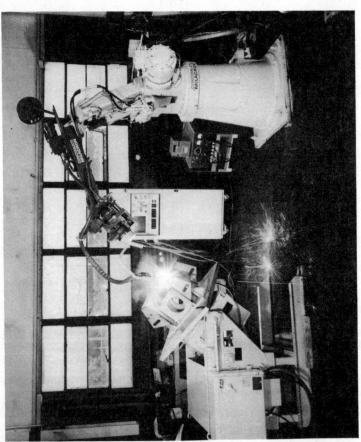

Fig. 4-2. The Cincinnati Milacron T³ industrial robot doing its job (courtesy Cincinnati Milacron).

type robots, perhaps with some modifications and improvements, will be doing assembly of intricate parts, moving various batches of materials from place to place and so on. They will in the final stages of an assembly line conduct the testing and evaluating of manufactured items and will accept or reject such items as necessary.

Examine Fig. 4-3, which shows the articulation of the arm, and its ancillary equipment. Notice that both electrical power units and hydraulic power units are used to obtain the necessary movements. You can see the hydraulic piston inside the elbow, for example, and you know that electric motors are used at the wrist and gripper because they furnish adequate power and are fast and small and relatively light weight. In Fig. 4-2, if you will refer back to it, you can see the large electric motor which is located at the shoulder position.

It is of some interest to examine more closely the wrist and gripper part of the arm as shown in Fig. 4-4. We see some details as to the number of motors used. Their size, and how the whole section is designed so that there can be little open space for dust and dirt to enter into the machinery is illustrated. Various types of grippers, the robotics word for fingers, can be attached to the disc.

On the application shown, one must realize that there must be a coordination between the arm and its companion, the holding and turning unit. There is this coordination, and it comes about by means of computerized control. Thus it is incumbent upon the computer to know, exactly, where in space the gripper is at any moment, and where the position of the work is. A programmer had to figure out just how the arm should move, that is, how much of a shoulder movement, how much of an elbow movement, how much wrist movement, was necessary to get the torch to the required points in space each time a move of that torch-tip was required. As you can easily see, there are countless movements of the arm itself and all its moving parts which might be used to change the position of the torch in space. But perhaps only one set of movements will give the end product, the required movement in the shortest time and in the most efficient manner. That kind of

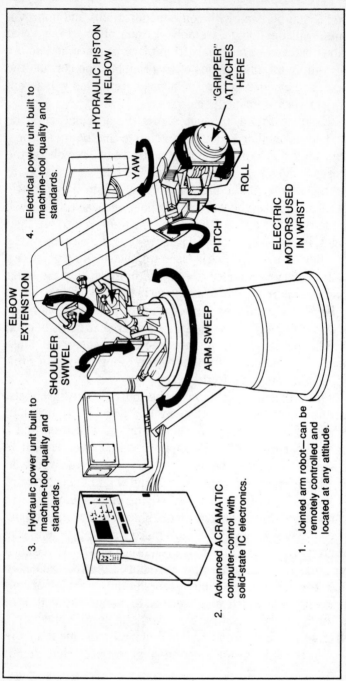

YAW

ROLL

"GRIPPER"
ATTACHES
HERE

PITCH

HYDRAULIC PISTON
IN ELBOW

4. Electrical power unit built to
machine-tool quality and
standards.

ELECTRIC MOTORS USED
IN WRIST

ELBOW
EXTENSTION

SHOULDER
SWIVEL

ARM SWEEP

3. Hydraulic power unit built to
machine-tool quality and
standards.

2. Advanced ACRAMATIC
computer-control with
solid-state IC electronics.

1. Jointed arm robot—can be
remotely controlled and
located at any attitude.

Fig. 4-3. Cincinnati Milacron T³ robot's arm movements and its ancillary equipment (courtesy Cincinnati Milacron).

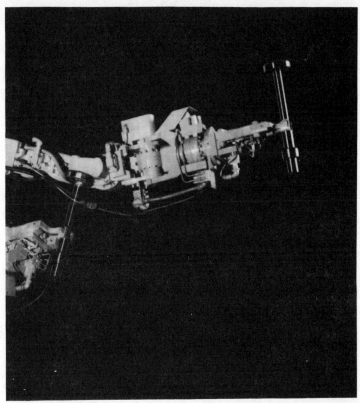

Fig. 4-4. The wrist and gripper of the Cincinnati Milacron T³ robot (courtesy Cincinnati Milacron).

movement must be the result of a good programmer's software development!

TEACHING THE CINCINNATI MILACRON T³ ROBOT

No, you do not have to be a programmer in order to operate this robot. But, what you will need is a small teaching control unit such as is shown in Fig. 4-5. You can examine the photograph and see the types of functions controlled by this unit. Essentially it will permit the movement of the arm gripper to any desired position, and it permits a check on the programming sequence by means of a replay, so that you know if the actions you have taken the arm through are really what you wanted it to do! Also, it is possible to have some programs stored on cassette tapes so that you can use them when appropriate. Change the arm movements with another tape and then later, if it is needed, you can go back to the first tape for the original movements.

Let us suppose that the job which we want to teach the robot is simply a pick and place operation. That is, it must move to an acceptance position. When the object that it is to move comes along it will grasp the object, lift it and move it around to another belt and then set it down, gently, and carefully, and then return to the acceptance position to wait for another object to come along. In this very simple example, we aren't worrying about how the robot identifies the objects, or if it is only to pick up certain objects from among many that might be on the belt. We are just assuming that at very precisely controlled intervals, along comes this thing which the grippers on the arm are positioned to intercept. We aren't even worried about the position of the thing. We assume it will be so positioned that the grippers will be able to engage it tightly in a proper manner and lift it without damage or slippage.

Let us examine the teaching unit of Fig. 4-5. Notice that we have some buttons which we can depress which will move the arm, the wrist, and the grippers, in a multitude of directions. At first we proceed slowly trying to make the arm swing around and set its gripper down on the belt in the proper position to intercept the thing. If we make mistakes, we can delete them and try again. Now we want to make the arm remember this segment of operation and so, through the proper button on the teaching device, we enable the computer to memorize those movements we have had the arm perform. We want to be sure. So we manually bring the arm around to some arbitrary position and we engage the section which causes a replay of our operation. The arm should move around to the desired position and open its grippers ready for the thing when it arrives. But, notice! The arm did not necessarily move in exactly the same manner we moved it! Now what? Is the arm beginning to think for itself?

No, the computer has taken all our little pieces of movement and examined them and then, using the software de-

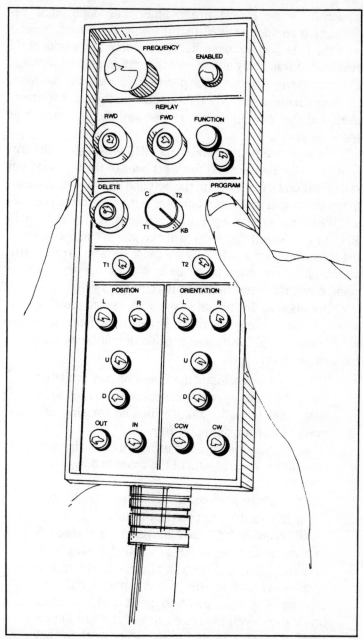

Fig. 4-5. The teaching unit for the Cincinnati Milacron T³ robot arm. Step-by-step movements are stored in the robot's minicomputer; the computer simplifies these directions and controls the arm movements.

signed by some expert programmer, has done some re-calculation and determined that the way we moved the arm isn't the best way to move it. So the computer works out a new spatial trajectory and moves the arm in its own manner. The arm moves to exactly the position we want it to move to. It orients itself in exactly the manner that we had it orient itself and the only difference is the way the arm moves in space to get there.

Now we need a thing and we need to slowly steer the arm so it picks up the item and moves it over to the new location and sets it down carefully on the belt and releases it. We take the arm through those movements, using the button control-ler. We then check our operation by a quick replay and if all is all right we commit that section of movement and operation to the computer memory. Finally we cause the arm to return to a starting position where it will lie in wait for the next thing to come down the input belt into its hungry jaws!

Because we don't want to make this seem too simple—although the teaching operation is relatively quite simple—we present the following description of motion as stated by the manufacturer:

> "During the teaching, the operator has available three different spatially coordinated motion sys-tems; that is, paths along which he (or she) can move the arm from point to point.
> *"Cylindrically coordinated motion* moves the arm around the vertical axis of the robot and in-or-out from that axis.
> *"Rectilinearly coordinated motion* moves the arm along (x,y, and/or z) coordinate lines.
> *"Hand-coordinated motion* moves the arm through as many as all three of the T³'s available axes (x, y, z, or roll, pitch, and yaw) simultaneously. Because all axes are coordinated, the arm moves only in a straight line from point to point and maintains orientation even though all three axes are in mo-tion. Hand coordinated movements of the robot arm can be compared best to the movement of the human hand which goes from place to place in the

shortest straight line path without any regard to any other spatial orientation of the body.

"All three of these motion systems are selected by pushbutton on the program entry terminal (teaching unit)."

DOING OTHER JOBS WITH
THE CINCINNATI MILACRON T³ ROBOT

The reason that this arm can do other things is that it can be equipped with sensors, and the arm is programmed from a computer directed base. This means, that like all computerized activities, software has been developed which makes it possible for the robot arm to understand human commands. Let's examine what the robot can do in a materials handling role. Figure 4-6 shows the arm stretched out at a low

Fig. 4-6. The robot arm selects a "thing" and picks it up and moves it (courtesy Cincinnati Milacron).

level to enable its grippers to grasp a thing. Next we can see how the arm contracts as it moves upward.

Note in the lower right corner the display unit of the computer cabinet. It is possible to look at the computerized program for this robot on that display screen, and programmers will be happy to know they can delete or add or even build a program using this display unit without having to actually move the arm at all. It might be somewhat similar to writing a program at home on your home computer. This screen also will display diagnostic information, should that be needed.

Finally, we mention that the reference point in this system is that spatial point where the gripper tool meets the workpiece. You have to know where that is, visually, if you program with the little control box, but not necessarily mathematically. The computer will mathematically find out and determine the space coordinates of that end point.

SUBROUTINE PROGRAMMING FOR A ROBOT

A subroutine for any computer generally means some pre-defined action which will be *called* at some point when an appropriate command or signal appears in the general program. In a way it is an aside action, and the overall operation of the computer is stopped while this side action is in effect. When it has been completed, the computer will then pick up where it left off in the main program and, possibly using the information from the side action, continue on its main program operation.

With a computer which controls a robot machine, the subroutine is sometimes called a branch action routine. This simply means that the movement of the robot arm will be made to follow something other than its normal movements when an appropriate activation signal causes it to go into the subroutine programming. We need to examine a situation which illustrates this.

A truly general-purpose industrial robot system must have some way of selecting or altering the normally programmed path and functions of its arm based on changes in the working environment in which it operates. The name given to this ability may vary from manufacturer to manufacturer, but the purpose of the operation will always be the same. When the robot's moving part reaches some point it will pause to see if there is another signal at that point. If that signal exists, then the robot will move into a subroutine. If, when the robot looks for that subroutine input signal, it finds that it does not exist, then the robot will continue its normal operations in normal sequence.

In Fig. 4-7 we see a diagram which shows a mainline program movement of a robot's arm, and a branch or subroutine break in that normal program movement. The exit into the subroutine movements occurs at C when an appropriate signal comes into the robot's input. Normally the arm will move on the trajectory A, B, C, D, E, and back to A. We can also assume that when the arm is not working it will be retracted to the position X, and that it must first extend from X to A before it goes around that mainline trajectory. We will imagine that the arm picks up something at A and moves it around to E, drops or positions it there, and then comes back to A to get another part.

Now let us see what could happen if an input signal comes at the proper time when the arm pauses at C. Suppose, just for an example, that the part picked up at A may need to have something else attached to it at F, G, and H before it is ready to be placed or positioned at E. The branching input signal then could mean that this part needs to go to all the stations. The main program trajectory line is changed to accommodate this required action. The arm goes into a branch trajectory or into a subroutine program at point C and exits that subroutine at D whence it is back on the mainline program trajectory once more. Notice that if the arm pauses at (C), and no input signal arrives, the arm will move on from C to D and will not go through the subroutine operation. It is important to realize that this is not a real-life example, but just a discussion and illustration to show what a branch program or a subroutine program is.

In practice there usually is a large number of these subroutine programs in operation. For example, if during inspection of a machined part some defect were found, the

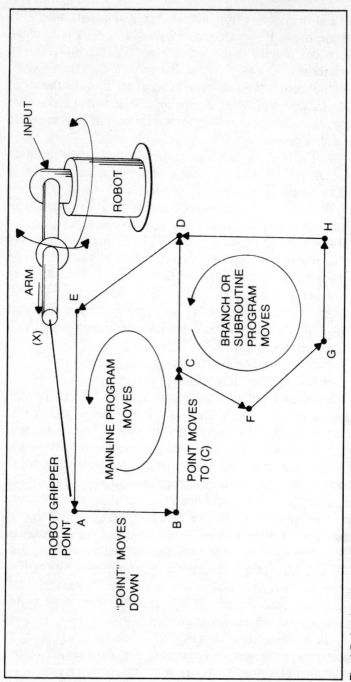

Fig. 4-7. A subroutine-controlled arm movement of a robot.

robot might branch to another part of the program that would dispose of that defective part, alert an operator to the fact that a defective part was there, and then the robot might just shut down operations (as specified by the subroutine) until the part was repaired or replaced or whatever.

The method discussed called standard branching is the simplest way to change the robot's path programming, but these standard branches are limited because each branch must be associated with one particular input signal to the robot's control system. This means that if there are X signals required there should be X lines to carry this information to the control system. It is possible to have fewer lines if an external decision-making element (a computer) is available which can look at all input signals simultaneously before specifying the branching or subroutine operations. With 8 lines there are 2^8 or 256 different conditions for branching that can be identified by the computer. In the Milacron Robot System the extension of the standard branch is called a conditional branch or subroutine.

THE COMPUTER OUTPUT FOR A CONDITIONAL-BRANCH SUBROUTINE

Let us see now what kind of computer output could result from having the computer look at all 8 input lines and the 256 possible conditions for branching. We are agreed, now, that the computer will look at all these possible inputs and will activate a subroutine or branch program if, and only if, the voltages present on the 8 lines represent a certain signal. For example, if the arm were programmed to branch into a subroutine at point B of Fig. 4-7, but the computer was told, "Branch B 1,-5,7,-8," then that subroutine would be entered into if, and only if, signal input 1 is on, *and* input signal 5 is off, *and* input signal ⌐ is on, *and* input signal 8 is off! If the computer looked at these particular signal inputs and found any of them different from the specification, then it would not command the subroutine branching at B and the arm would continue on its normal trajectory to C. Notice that it is possible to express the robot's operation completely in Boolean Algebraic expressions.

BRIEF BOOLEAN ALGEBRA REFRESHER

Boolean Algebra is called the algebra or logic. It was invented by George Boole, mathematician and logistician, in 1847. This is the algebra of automatic control and switching devices in today's world. It permits one to present switching circuits with algebraic symbols and then to manipulate those symbols according to the laws of algebra so that one can find the simplest manner in which to accomplish complex switching and operation of various electronics and electrical circuits.

Boolean Algebra deals with truths and non-truths that relate well to switching devices that are either on or off. Digital devices use "states" to represent one or zero which is roughly equivalent to a switch being on or off. Therefore, Boolean Algebra is well suited for digital work. Let us apply this concept to a simple circuit.

Let us say we have a circuit with a switch and a light in it. We can say several things about that circuit:

(A) The circuit is on
(B) The light is on

Algebraically we can use the letter symbols of these statements to indicate and express the condition of the circuit and the light. For example:

1. A.B (meaning A times B) means that the circuit is on AND the light is on.

2. A+B meaning either the circuit is on OR the light is on.

Finally we can write:

3. $A \cdot \overline{B}$ where the bar over the B means NOT or a negative. Thus, the circuit is on and NOT the light.

These are the three basic functions of Boolean algebra, they are the AND, the OR, and the NOT functions.

When this kind of algebra is applied to circuitry, it is common to use the multiplication function for series connections, see Fig. 4-8A. For parallel connections one would use addition (Fig. 4-8B), and a negative situation is at Fig. 4-8C. It

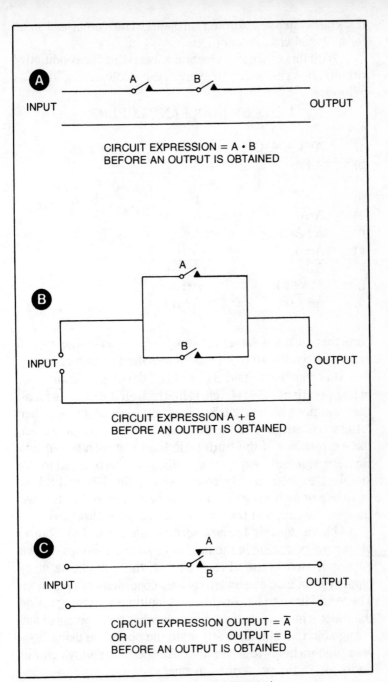

Fig. 4-8. Some Boolean algebra expressions illustrated.

is common to use a zero for an open circuit output and a one for a closed circuit condition.

With this concept of algebraically stating the condition of circuits, it is customary to express the Boolean conditions as follows:

LAWS OF BOOLEAN ALGEBRA

(a)	$A \cdot B + A \cdot C$	$= A(B+C)$	($\cdot$ = and)
(b)	$A+A$	$= A$	($+$ = or)
(c)	$A \cdot A$	$= A$	($\overline{A}$ = NOTA)
(d)	$A+\overline{A}$	$= 1$	(Key)
(e)	$A \cdot \overline{A}$	$= 0$	
(f)	$A+A \cdot B$	$= A$	
(g)	$\overline{A}+\overline{B}$	$=(\overline{A \cdot B})$	
(h)	$\overline{AB}$	$=(\overline{A}+\overline{B})$	
(i)	$A(\overline{A}+B)$	$= A \cdot B$	
(j)	$A+\overline{A} \cdot B$	$= A+B$	

Note that the dot is sometimes dropped, just as in algebra, and order is from left to right, unless specified by parentheses.

It is common to take a circuit and develop a Truth Table of its operations. This is a chart in which all the outputs for all the conditions of the switches are stated, one at a time. Then letters are assigned to the switches, and, as you have seen, the expressions of the truth table lines can then be put into Boolean algebraic expressions which can be reduced to the simplest expression. By re-converting the letters back to switches or switching circuits, a person is then able to draw the best circuit, and the simplest circuit, for that device.

One can quickly see how such a mathematical tool can be of tremendous value in the field of advanced robotics. As we have indicated in the Milacron robot, there might be many inputs to the machine under various conditions of operation. There will have to be conditions of multiple switching inside the robot's mechanism to make it do what the inputs call for. Being able to mathematically state the operation using Boolean algebra helps to arrive at a cost-effective unit which will do its job in the prescribed manner.

MORE ABOUT ROBOT BRANCHING OPERATIONS

To illustrate one use of conditional branches, imagine that you have a material handling situation in which a supervisory computer is being used, and this computer is also being used for other automation functions. Remember that the number of lines required for a given robot's operation can be stated in powers of 2, then, if the robot machine requires 16 signal inputs, the robot will require only four lines to convey to it all the information it needs to get the proper branch program into operation.

In many applications of a robot machine in industry it is desirable for the robot to have a method for creating a branch operation that can be used at a number of points in the robot cycle; a branch that can effect its movement sequence, regardless of the physical location of the robot arm, when the branch operation is requested. For example, when the branch operation affects the robot's gripper position and/or its wrist orientation.

Some kinds of applications require another kind of offset branch, where the branch can be entered, not only at specifically programmed points, but anywhere or anytime in the normal cycle of operation. When a proper input signal is received by the arm-control unit the signal can cause the robot to enter into an interrupt offset branch operation. One example is that illustrated in Fig. 4-9. In this illustration one can see that the arm is caused to move on trajectory 0B01-0B02 when the interrupt signal is received at point X on the normal trajectory from M005 to M001. Notice also that at each position M002 and M003 and M005, the robot's arm will go into the type of branching operation specified by the 0A numbers. Notice that the type of movement for each of the M points is exactly the same when the branching occurs in the 0A series, as shown in this illustration. So the same subroutine called for at point M002, can be called for at M003 and M005.

The question as to what an interrupt branching operation might consist of may be answered by considering that it could be a twisting, jerking, series of movements of the robot's arm

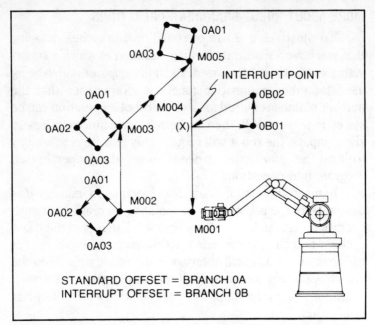

Fig. 4-9. Offset branching of a robot arm (courtesy Cincinnati Milacron).

which would cause a welding gun tip to be loosened from a stuck position on the work piece. A sensor of some type would have to monitor the gripper movement, and in event it becomes stuck or does not move along smoothly as it should, then this sensor generates the signal which causes a special routine to be implemented immediately by the computer. That, of course, in turn, activates the arm into the special motions described.

Extending this concept one can realize that special routines might be required for many applications. If the robot's arm handled the moving of objects which normally are spaced a given distance apart, and timed precisely on arrival at a given position, then if something disturbs the flow, a sensor would generate, perhaps, a wait or delay signal to the arm.

THE ROBOT'S SEARCHING FUNCTION

In addition to altering its cycle by entering branches, a robot system such as the Milacron can adjust the position of data points within an existing cycle, based on changes in external equipment and workpieces. In a way, this might be called adaptive control where the current condition of something specifies what the future operation will be.

One example of how this type of adaptive control might be used is in stacking, or handling operations from a stack, where the items may have a variable thickness, or the stack may vary in thickness or height. In this case you can readily realize that the arm must progress from some fixed-level starting point, where it begins to "pick up" the parts, and descend to lower and lower levels each time a part or work piece is moved out of the stack. The control operation in this type of movement is called a search function.

The data comprising a complete branch sequence in the robot control can be rewritten with data from the external computer system. Thus, if the position or orientation of a workpiece varies, the robot's movements and functions can be adjusted by the external device so that it will properly pick up or perform the required operations on the workpiece. If an emergency situation is detected, a branch can be replaced with other branch data which will enable the machine to handle this specific emergency condition. The speed at which this data can be transmitted to the robot control is fast enough to complete the communication in a fraction of a second, so this type of branching has been and is called adaptive branching.

MORE ON THE SEARCH FUNCTION

The search option on Cincinnati Milacron's robot can be used in stacking or unstacking applications. In teaching the robot the task, two points, one above the stack and one at the bottom, are programmed. Then a tactile sensor on the robot's hand is used to send an interrupt signal to the robot control system whenever the hand encounters an object in its path, as it moves between the two programmed points. The interrupt signal causes arm motion to cease immediately. Then the function associated with the subroutine is executed. The arm then proceeds to the next point in the main program which may be back to the original starting point.

In an example, imagine that the robot arm is to pick up some automobile windshields from a stack of them, and place them one at a time on a conveyor belt, off to one side of the stack. Recall that only the starting point above the stack and the bottom point, at the bottom of the stack, are necessary to be known by the robot's control machine. Thus, when the arm goes from the start position down the stack, it stops itself when its sensor contacts a windshield. The gripper then grips that unit, and immediately moves it along a prescribed path to the conveyor belt and deposits it there. Then the arm goes back to the start position and resumes the operation all over again. It is not necessary for the program to include the specific pieces of work on their positions, the sensor on the robot's arm takes care of that. Only when the robot's arm reaches the second point, or stop position—meaning it is at the bottom of the stack—will the arm move to some rest position and stop moving.

Of course we realize the possibility of interrupting a normal branching operation with an interrupt signal. This signal might cause the arm to skip some movements and move on to others, or go into an entirely new routine. Of course any routine must have been taught to the computer by moving the arm, with the teaching unit, or must be specified by signals on a cassette or other similar type tape.

COMPUTER-USE SUMMARY

So we have learned that a computer can direct a robot—which, of course you already knew—and we now have some concept of how the computer does this. The output of the computer will be a series of signals to a bus which might have, in our case, say, eight lines or terminals. When various keys are pressed on the computer keyboard, the closing of the key-switch contact will cause the computer, through its internal circuitry, to energize some combinations of these terminals. Since a terminal status can be specified by being either on or off—a two state condition—we can then consider the permutations and combinations of eight things taken in two conditions and we find that there are 256 possible states of output, each being completely different. Of course if we go to a 16 line or terminal type output on the I/O board, then we will have 2^{16} unique states for control. We leave it to you to determine the numerical value of this possibility.

In order to control the higher values of voltage and current necessary to really operate the control system of a robot, some isolation condition must exist between the computer and the controlled circuitry. This can be accomplished by using optoelectronic circuits, or reed relays, or some kind of similar devices. The computer output then operates these devices and they, in turn, will operate the higher current and voltage control lines.

When sensor input is required to the computer, these signals come from sensors attached to the robot's positioning and moving elements or from scanning elements such as TV microwave, sonic, or thresh-hold switches. Sensors such as piezo crystals generate a current proportional to the force on their containment cases. This voltage then must be digitized as it originates in an analog form from the sensor. In other words, it exists as a continuous level of current or voltage and it must go to some circuit which will convert it into a computer word. As a very simple example suppose that the amount level of the sample is one volt, then the word might be 00000001 for an eight bit word. If the voltage level is two volts, the word might be 00000011 and so on. In any event a particular series of off-on pulses or bits will now represent the sensor level signal to the computer.

The computer can work with this. It routes the pulses to various addresses which in turn release new signals to the I/O board which then passes them on to the control circuitry for the robot's moving elements. Notice how the computer might get a particular feedback signal from a sensor, and when this signal goes to the memory banks it might cause a subroutine program to be initiated. The subroutine will be unique and special and non-duplicable by any other input signal. So, anytime it is necessary for the robot to move in accord with this particular subroutine program, that particular signal of ones and zeros must be received by the computer.

SOME CONSIDERATIONS OF ROBOT DESIGN

The robot is a machine and can be a very complex machine if the situation and conditions warrant this type of development. The robot falls into a category of machines which are usually studied under the heading of feedback servomechanisms. This is really a fascinating subject and it embodies the best application of such mathematical concepts as the LaPlace Transforms and Fourier Transforms to express mathematically how such machines will operate and react to various input or operating conditions. When designing such a robotics system one takes into account such things as:

The system environment
Noise
Nonlinearities which can exist in the system
Component limitations
Undesired disturbances
Size and weight and space and power requirements
The random inputs which may occur.

If you get into the design of such devices, then you should be familiar with such mathematical techniques as the *Maximum Principle* of Pontryagin, Bellman's *Principle of Optimization*, and Wiener's *Least Squares Optimization*. Some suggest that classical calculus variations of these should be understood. These mathematics enable a person or designer to put down on paper the design of a system and the analyze its operation before building it. Thus when it is built, it can be expected to operate in a given manner and within a given type of specification range. Computers enable one to solve the most complex of problems easily, so all one has to know is what mathematics to use and how to set up the problem for computer solution and what the computer solution actually means!

We have mentioned adaptive control in connection with a robotics system. It might be important to examine this concept just a bit more in detail. Adaptive control means, simply, that the machine will adapt its operation to a change in either its external or internal environment. A self adaptive control system is one which has the ability to change its operations through a process of measurement and evaluation and adjust itself to a new set of conditions in which it operates.

What a robot design engineer is always looking for is the optimum response or operation of the system. That is, the system must not work too fast or too slowly, but at the right speed for the process or job at hand. If there is some change in its environment—say that this might be the speed at which parts are sent to it—then it should be able to adapt itself to the new speed of arrival of parts and continue on merrily and happily without any problems.

To make a system which does things at the right speed, as an example, we make an assumption that we know what the right speed is. We then program this speed into the machine and provide it with sensors which can determine at what speed it is operating. Then, inside the machine, some computing element compares its speed at present to the speed which we say is ideal, and the machine can then take the difference between these two and uses that as a control signal to adjust the speed of operation. Of course, you might imagine that the speed signal we send into the machine when the operation starts is just a priming signal. Once the mechanism is in operation, this signal is constantly adjusted by adaptive circuitry so that the optimum or best response for that situation is accomplished.

In the advanced world of robotics design you make use of sampled data systems as well as adaptive control systems.

SOME CONSIDERATIONS IN PROGRAMMING A MOBILE ROBOT

It will be beneficial to consider some aspects of programming a robot that can move about. We know that if a sensor senses something which might call for some movement of the robot, what that sensor really does is to cause a signal to be generated which then calls a subroutine that causes the necessary electrical power to be applied in such a way that the movements are accomplished. Let us consider how we might program a robot to move in a given way in a given area.

As a first consideration we are faced with the basic question, "Where is the robot? What are the robot's X and Y coordinates at any second? What reference plane are those coordinates on? Luckily for us, some manufacturers have developed what is called a stepping motor. It can help us answer the basic question.

As you know a stepping motor is one which revolves a given amount when supplied with a pulse of electric current. That amount might be five degrees or it might be twenty five degrees. If pulses are supplied in a continuous train, the armature of such a motor revolves in a continuous manner, and if the pulse rate is fast enough we will not be aware that there is any difference between the rotation of this type motor's armature and the armature of any other motor. Both revolve smoothly and quickly, but, there is a difference, and an important one to us.

With a stepping motor we can govern distance traveled or arm movement quite precisely. If we have a motor which turns five degrees with each applied pulse, due to the way it is built there must be a space (no-pulse) following each pulse, or it won't work right. Divide 5 degrees into 360 to get the portion of a turn each pulse represents. This number times the circumference of the motor wheel is how far that wheel will move on a floor with each pulse. Next we make a small diagram of the movement our mobile robot should follow as shown in Fig. 4-10.

What we do is to program a number into the computer memory, for this example the number 11. Next we program the computer so that when this number has been counted down a subroutine will activate which will cause the steering wheels to turn as shown through the angle alpha. The computer knows that this has been done by monitoring a voltage proportional to the angle the steering-drive wheels make with the robot's frame, which can come from a potentiometer mounted on the steering section of the robot. Now a second drive subroutine count-down section begins and at the same time the steering wheels will equalize in speed turning straight ahead almost immediately after the robot starts moving. It will move straight along the X path for 8 pulses, turn again thru the pre-set angle and continue 4 pulses to the stop position. If you try to make such a robot, some experimenta-

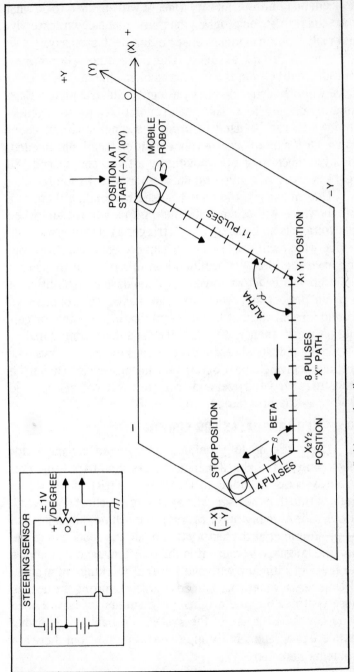

Fig. 4-10. Map of pulse drive wheel controlled by a subroutine.

137

tion will be required, but this concept will enable it to do what you want it to, *if* you program the pulse count-down correctly, and make the turn-angle sensor equipment properly.

This is just an example. There are other alternatives. One alternative would be to put a small magnet on the face of a drive wheel which is driven by a regular electric motor. Each time the drive wheel makes one revolution a pulse is generated which can be used to make a computer circuit count down. Or if a more precise robot position location is needed, then two such magnets, moving by a fixed coil spaced 180 degrees apart will give two pulses per wheel rotation.

One might use the wheel pulses by sending them to a circuit which will accumulate them into a voltage and when this voltage is high enough, it can trigger a relay or some such device which will cause other circuits to energize. Now that you have the idea, you can build on this concept to achieve something satisfactory to your own needs and capabilities. It is a fun project and when you can make your robot move around a room without hitting anything, or stopping, or getting off track, then you will really have done something!

In the industrial sense, the previous example shows how a robot's arm might be programmed during a cycle. Then if the program is repeated over and over, the cycle will be repeated as long as the program is run.

COMPUTER LANGUAGES FOR ROBOTICS

There are no fully defined, recognized languages for robot control. Individual manufacturers tend to put together whole systems, including CPU, and define each movement in a subroutine that the user calls by typing in a short code word. These subroutines are, of course, in machine language, and the programmer at the factory that made the robot had to use machine language because it is the only language that allows the level of intimacy with the internal workings of a given CPU necessary for programmed control. Limiting the user to those routines that the original programmer foresaw a need for is not the best use of the system's capability. Allowing complete programmability appeared preferable, but there are problems here too.

First, many programmers lack the combined hardware and software knowledge needed to do control programming. Second, the company is understandably concerned about disseminating too much information dealing with the internal control of a machine that they have invested years of design work and money in. Third, most companies installing such a system would probably not want to hire a programmer just to do one job (though this is changing) and a short term contract with a specialist is pretty expensive. There is a solution though, and most companies will probably be going to this method eventually, though today most are still using the first method discussed.

There are high-level languages that are very easy to program in but do not compile down to very efficient machine code. Most of them have the capability of linking subroutines that are already in machine language to a program being compiled. So the manufacturer writes a group of very efficient routines for directly controlling the robot, and supplies them as a library of added functions to a language like BASIC or FORTRAN.

There have been some studies of robotics conducted at Stanford Research Institute in which FORTRAN has been used. In one such study a mobile robot was programmed to move about a floor space and was asked to move certain objects, by pushing them, into various positions in that space. This study was conducted for the Advanced Projects Agency of the Rome Development Center, Griffith Airforce Base, New York. It is interesting to examine a portion of their report.

> "The robot system is capable of executing a number of functions that vary in complexity from the simple ability to turn the drive wheels a certain number of steps, to the ability to collect a number of boxes by pushing them to a common area of the room.
>
> "Of the functions mentioned the simplest are certain primitive assembly language routines for moving the wheels and tilting the head (TV), reading a TV picture and so on.

"Two examples are MOVE and TURN. MOVE causes the robot to move in a straight line by turning both wheels in unison, and TURN causes the robot to rotate about its center by turning the drive wheels in opposite directions. The arguments of MOVE and TURN are the number of steps that the drive wheels are to turn (each step resulting in a robot motion of 1/32nd inch) and status arguments which allow inquiries to be made about whether the function has been completed or not.

"Once begun, the execution of any function either proceeds until it is completed in its normal manner, or until it is halted by one of a number of abnormal circumstances such as the robot bumping into unexpected objects, overload conditions, resources exhaustion, and so on. Under normal operation if the execution of the MOVE command results in a bump, motion is stopped automatically by a special mechanism on the vehicle. This mechanism can be overridden by a special instruction from the computer, however, to enable the robot to push objects.

"The problem solving systems for MOVE and TURN are trivial. They need only to calculate what signals shall be sent to registers associated with the motors in order to complete the desired number of steps.

"At a level just above MOVE and TURN is a function whose execution causes the vehicle to travel to a point specified by a pair of X,Y coordinates. This function is implemented in the FORTRAN routine called LEG. The model used by LEG contains information about the robot's present X,Y location and heading, relative to a given coordinate system, and information about how far the robot travels for each step applied to the stepping motors."

...   .

"Ascending one more level in the system we find a group of FORTRAN two-letter routines whose execution can be initiated from the control machine keyboard.

The action unit system now ceases to be strictly hierarchical (arranged in a graded series or sequence) since, now, some of the two letter commands can cause other commands to be executed. This means a branching type subroutine operation.

"In this system, developed and studied at SRI, one two-letter command (EX) takes as an argument a series of X,Y coordinate positions in the robot's world. Execution of EX causes the robot to travel from its present position directly to the first point in the sequence, thence, directly to the second point in the sequence, etc., until the robot reaches its final destination as programmed. Another two-letter command (PI) causes a picture to be read after a TV camera is aimed at a specific point on the floor to see if there is an object there or not. The ability to travel by the shortest route to a specified goal position along a path calculated to avoid obstacles, is programmed by the two-letter command TE. Execution of this command requires that the computer solve the path coordinates using a special problem solving system which is activated by the two-letter command PL."

... ...

"As SRI ascended to higher functions of operation, the required problem solving process had to be more powerful and general. They wanted this robot to be able to perform tasks which, quite possibly, would require complex, logical deductions. What was needed then for this type of problem solving was a general language in which problems could be stated, and a powerful search strategy with which to find solutions to these problems. They chose the language of first order predicate calculus to state the problems to the robot. The problems were then solved by an adaptation of a question answering system based on resolution theorem-proving methods. One example of such a problem for the robot to consider, stated in plain English is, "There exists a situation s and a place p such that three objects called OB1,OB2 and OB3 are all at place p in situation s."

Now the problem as the robot sees this is that the three objects may not all be at place p, nor are they in situation s. So the robot must move them until this problem statement is satisfied. Notice that the command is not given to the robot to find OB1 and push it to place s, find OB2 and push it to place s_1 and find OB3 and push it to place s_2.

In this case one would suspect that the robot would move its camera eyes around until it could identify OB1 either by color or shape or both, and then track it down and move it to place s where s has been indicated to the computer in the form of X,Y coordinates. The proof of the action would be that the robot, from a position x_1, y_1 would look at the space s and if it there found the form of OB1, it would then look around for OB2, etc.

We have included this information on this study because it indicates some methods of programming robots, and the use of a two-letter FORTRAN language. In this study, the SRI group also used LISP language. It is stated that

"... Another LISP program enables commands stated in simple English to be executed. It also accepts simple English statements about the environment and translates them into predicate calculus statements to be stored as axioms. The English processing by this program is based on work by L.S. Coles: *"An On-Line Question-Answering System with Natural Language and Pictorial Input"*. (Proceedings 1968 Conference of ACM, Las Vegas, Nevada, August, 1968.)

In a survey of the languages which have been used in robotics control in the past, and in work on artificial intelligence, we find such descriptive names as PLANNER, CONNIVER, SAIL, STRIPS and QA4.

THE MULTIARMED INDUSTRIAL ROBOT

If we consider the manner in which a Homo sapien accomplishes a given physical task, such as the assembly of one part to another using a bolt, a nut, and perhaps some washers, we find that not only are two hands and fingers used, but also, in some cases the fore-arms and even an immobile device such as a vice, which holds the work to which the piece is to be bolted. In a manner of viewing this situation one can say that the human has provided himself with another hand or two in order to hold everything together so that the bolting process can be accomplished. In the simplest sense, the two parts are held by a vice, one hand inserts and manipulates the bolt, and the other hand holds and/or manipulates the nut onto the bolt, after first placing the washers in position.

From this example then, we should be aware that a system of assembly in an industrial situation might involve a robot machine with three or four arms. Actually the device which positions the workpiece can also be called a part—even if it is a remote part—of that particular robot system, and so we now have to consider that a robot may not be an integrated entity all in one frame or piece, but may consist of many parts which are tied together by means of command system. In the operation of such a unit, coordination among the various pieces is affected by a computer which may have as an input, the outputs of many smaller computers, and its output may become the input for the smaller computers, thus effecting the control and timing of all elements of the system.

It seems that we might be approaching the concept of HAL in the story *2001—A Space Odyssey*. The series of computers form the nerve networks and sub-brains associated with a complete automated plant's operation. This can be especially true if we consider that everything done in a given plant is to produce a product at the output, and there must be coordination and timing and productivity of the various parts as well as assembly stages and testing stages in order to finally send the finished product off the assembly line at the plant's output. Under this concept we do not have a series of robots here and there, but in effect, a gigantic robot with all the complexity needed to start with raw materials, and produce finished products on a constant 24 hour operation basis. Even suppliers, distributors, wholesalers, and retail outlets might provide demand inputs to the controlling computer. The concept becomes mind-boggling!

TESTS, ADJUSTMENTS AND REPAIRS

Even now we have computers with diagnostic programs. One can envision in the robotics systems of the future that the machine itself, when it gets a signal from some source that something is not going exactly right, will stop, initiate its own diagnostics program to find out what is wrong, and then, using visual and audible communications, tell its human supervisor verbally to fix the bolt on number 4, left hand, third finger digit. In all the current studies, at least to date, one doesn't find robots capable of repairing themselves without human help.

Also, these robotics systems must have diagnostics programs written, and fed to it by human hands or voices. The need for people in such systems is still current and in evidence. A diagnostic program is really just a check-through of the operation to see if all the movements and operations required are accomplished with the normal input signals. If, at any stage, the robot's response is not correct, then a signal will inform the supervisor that some adjustment, testing, or repairs are needed.

Programming the
Prab Versatran Robot

5

Prab Conveyors, Inc. have developed the Versatran robot which may be operated by a microcomputer, and is similar, in its learning operation, to the Milacron which we have just examined. Prab robots are also designed such that they can be programmed with a type of memory drum control, that is, it is possible to insert pegs into a revolving drum in such a way that these pegs will close electrical circuits, and open them, in the proper time sequencing so that various desired operations of this type robot can be accomplished. Also, this type robot can be controlled by limit switches, switches which open circuits at extremes of movement, or when certain velocities of movement are reached. They also can cause an arm to stop moving by limiting it physically—a mechanical stopping method.

We have mentioned the palletizing operation of an industrial robot. Look at such an operation, using a Prab Versatran robot in Fig. 5-1. We see the gripper holding a bale, in the upper left side of the illustration, the arm will move to deposit it onto the stack in the proper orientation for stability of the stack. At the extreme right of the illustration you can see the control panel which governs the movement of the Versatran arm. We might not necessarily associate this automated machine with the concept of a robot, such as we imagine from science fiction tales, but it is a robot in every sense except physical appearance.

We can view some other applications of this type robot in Fig. 5-2, where it has an arm moving from a shoulder section to make plastic automobile parts. Also, we can obtain another view of this machine in operation in Fig. 5-3, where it is serving a five-station machining system which produces 80# ring gears.

Each of these units has a microcomputer type brain that is programmed by teaching the arm the various positions it must reach, and the sequence in which it must reach them. This is done by entering these desired positions, in the proper order, on the microcomputer keyboard. The software then causes the arm to move properly and in the most efficient manner to accomplish the desired objective.

Each machine must have feedback signals so that the computer can determine when a given limit of movement is reached. To do this, the machines are equipped with limit switches, potentiometers, various types of encoders (devices which can tell the computer how much a shaft has turned, perhaps by generating a pulse for each fractional rotational movement) and resolvers (a selsyn type of unit that generates an ac signal whose phase change is directly proportional to the amount the shaft has turned).

Programs for this type robot are written according to the manual data input, and can be recorded on magnetic tape cassettes. Other computers may be used with these robots using a special interfacing unit. The users may write their own programs for the machine. The Prab Versatran robot can move to any of 4,000 points in space.

CONCERNING ACCURACY OF A POSITIONING SERVO-SYSTEM

If we consider that a positioning system depends upon feedback as the control element in determining the system error or operative voltage, then we must look at those points from which feedback occurs. Recall that the operating or error voltage or signal for a positioning servo is derived basically from a comparison of an input signal command to the

Fig. 5-1. Prab Versatron robot palletizes 80 lb. bales of rubber (courtesy Prab Conveyors).

ROBOT ARM

SHOULDER
SECTION

Fig. 5-2. Prab robot in a plastics auto-parts production plant (courtesy Prab Conveyors).

EXTENDED ARM POSITIVE
PART FOR MACHINING

XLO Michigan Tool

Fig. 5-3. A Prab robot in a ring-gear machining operation (courtesy Prab Conveyors).

149

feedback signal which is generated from the position of the output member in question. If we assume that the command signal is simply a reference, then it is the magnitude of the feedback signal and its polarity or phase which will govern the effect of the two signal comparison, and thus govern the size of the error signal. Since the system power unit operates on the difference (error) signal, then the closer the output arm comes to the commanded position, the smaller the signal becomes. At some point this signal becomes so small that it will not cause power to be applied to move the arm further. Where is the arm, or end appendage at that moment? Is it *precisely* where the command calls for, or is it within some tolerance, perhaps in the thousandths or even ten-thousandths of an inch from that position? It cannot be perfectly positioned, thus it is within some tolerance.

Now, to gain a full appreciation for the complexity of these systems, note that when moving the end appendage, the system must reach many positioning points. First, for example, the arm must rotate to a given azimuth and stop there within the tolerance accuracy. Next the arm must extend to a given distance and stop there within a specification accuracy. The end appendage may have to rotate about two axes within given accuracy specifications, and finally that end appendage may have to advance a given distance to within a specified tolerance. One notes, then, that errors in one element may effect the errors in another element. The errors may become cumulative, affecting the overall precision.

It has been argued that because the end device generates a feedback signal of its own that all other errors in the system's moving parts will be compensated for. However, there is still velocity and acceleration control that must also be within tolerance.

If a robot were used where it puts a bolt through a small tolerance hole, then the tolerance or accuracy of the robot's positioning of the bolt must be within the hole tolerance or better! In a case where the robot simply grasps something from a conveyor belt, such tight precision may not be necessary. The point is, it is the overall accuracy of the robotics system which, in large part, controls its cost! The more precise it must be the higher its cost.

PROGRAMMING AN INDUSTRIAL ROBOT: THE PRAB VERSATRAN

Through the courtesy of Prab Conveyors we will now examine their Versatran robot to learn how it is programmed.

General Information

Versatran robots combine electronic, mechanical and hydraulic action with flexible programming to provide general purpose machines capable of performing highly accurate and precisely controlled work functions. The object of the Versatran System is to drive the robot arm through a combination of motions and directions to perform a specific task.

The Versatran System is classified as a point-to-point automatic system. In a point-to-point system, program commands define specific positions or points in space to which the robot arm is to travel. The robot arm will move simultaneously in all axes (directions) requiring change to reach the defined position. In this way the most direct route will be taken to the new position. In the basic 3-Axis Versatran System, the Horizontal, Vertical, and Swing axes are provided. Optional wrist motions, including Rotate, Sweep, and Yaw axes, can be added to the tooling end of the robot arm to extend its work capability. Also a Traverse motion can be added when it is desirable to have the robot move from one location to another in performing its operation.

Instruction Codes. The model 600 Control Unit contains a Program Teach Panel which is used by the operator (programmer) to construct a sequence of instructions which direct the robot to perform a specific task.

A name is assigned to each instruction which in some way describes what the instruction does. Each instruction is made up of two (2) parts: The Function Command, which informs the Control Unit of the general task to be done; and The Function Value, which identifies specifically what action should take place. Coding the instruction is simply the act of combining a function command with a function value to produce a definite action.

Programming Rules. Programming or "teaching" the Control Unit is simply the act of establishing a sequence of coded instructions which directs the Control Unit to perform a particular task.

Programming is only as successful as the operators ability to understand the fundamental activities of the task to be performed. A good, sound structuring of a program is advisable to keep programs simple, clear and complete. The following five (5) step procedure to successful programming should be employed when preparing a program:

RULE 1. Define the Operation: This is a brief description of the operation. The description should be prepared in a way which tends to identify the activities of the task.

RULE 2. Chart the Sequence of Events: This is a detailed description of the specific operation to be done. Each step of the operation should be listed in the sequence of which it is to be performed. (Depending on the complexities of the operation, it may be necessary to prepare sketches or flow charts in order to identify each step).

RULE 3. Code the Program: From the detailed description, prepare a list of coded instructions in the sequence in which they are to be "taught". This list of instructions is referred to as a Source Program from which the Control Unit operates.

RULE 4. Teach the Program: This is the ENTRY of the Source Program into the Control Unit in the sequence established by the Source Program.

RULE 5. Check the Program: Once the program has been entered into the Control Unit, a review of the program instructions is advisable in order to assure that all instructions are in their proper sequence and the CONTROL UNIT will execute them properly.

Program Teach Panel. The Program Teach Panel is divided into five (5) basic categories. Prior to preparing any program, it is necessary for the operator to thoroughly understand the purpose of each control.

(a) Operational Controls: These controls are used primarily for the operator to execute certain operational assignments such as teach, run program, erase program, etc.

(b) Manual Controls: These controls are used primarily for establishing, recording, and modifying the robot arm positions.

(c) Program Function Controls: These controls are used primarily for entering a set of coded instructions in a sequence that establishes a particular task for the robot to perform.

(d) Program Review Controls: These controls are used to execute each instruction, one at a time, in order to assure the program fulfills the intended task or to begin the program at a sequence other than with the first instruction.

(e) Program Editing Controls: These controls are used to alter or change the program or program sequence when corrections are necessary.

Operational Controls

The Operational Controls are used to assign certain responsibilities to the Control Unit.

These Controls consist of the Mode Select Key, Program Number Control, Program Erase, Lamp Test, and Cassette or Hand Held Teach Unit operation.

Mode Select Key Switch. The Mode Select is a turn-key operated switch used to select the Auto-Run mode of operation, Teach mode of operation, Test Run mode of operation, or the Single Step mode of operation.

(a) Auto-Run: The Auto-Run position is the normal mode of operation when the robot is used to perform a specific task. Placing the key in this position and pressing the Run Program button on the Operator Control Panel causes the robot to proceed with the task (program) identified by the Program Number Indicator.

(b) Teach: The TEACH position is used for preparing new programs or for checking and modifying existing programs. The new programs can either be manually entered with the use of the Manual and Function Controls, or electronically entered through the use of the Tape Cassette unit.

(c) Single Step: The Single Step position is used when it is desirable to review each instruction in sequence.

(d) Test Run: The Test Run position is used to check the movement and positions of the Robot Arm as established by the program instruction sequence. The Robot Arm is allowed to complete its cycle without interruption which may normally be required by the program.

Program Number Request. Selecting the PROG NUMBER Control will allow the operator to request any of the programs (within memory) to be run. THERE ARE 64 POSSIBLE PROGRAMS WHICH MAY BE SELECTED.

PROCEDURE:

(a) Place the MODE SELECT key in the desired position.

(b) Press the PROG NUMBER Control on the program teach Panel. (The lamp will light indicating the request).

(c) Select the desired program number on the function keyboard (1-64).

(d) Press the ENTER pushbutton under the keyboard (the PROG NUMBER indicator lamp will go off and the Program Number selected will be displayed in the PROGRAM NUMBER INDICATOR).

The Program is now available to be modified or run.

If the program selected is a valid program, the Mode Select key can be placed in the Auto-Run position to operate the program. If there is nothing in the selected program, turning the Mode Select key to any of the run modes will result in a code "130" display.

—CAUTION—

When operating the JOG control observe that the Robot Arm travels a path which is free of any obstructions. If the Robot Arm cannot be jogged to position without striking an object, release the JOG pushbutton and operate the manual axis request and axis direction controls to move the robot arm around the obstruction, then continue operating the JOG control until the robot arm arrives in position.

Program Erase. The Program Erase control is used to clear the program identified by the Program Number Indicator from memory. Once the program is erased all instructions and recorded positions are cancelled from the program.

<u>PROCEDURE:</u>
(a) Place the Mode Select key in the TEACH position.
(b) Select the Program to be erased from memory by using the Program Number Request Procedure.
(c) Press three (3) times in sequence: PROGRAM ERASE then ENTER. (Three times are required to prevent accidental erasures.

Lamp Test. The Lamp Test is a Control Unit exercise which scans all of the functions of the Unit. In this way a self-test operation can be performed. Each control button on the Program Teach Panel which normally lights will be lighted in sequence, beginning with the PROG NUMBER control at the upper left of the panel and continue to the right. At the same time the Digital Indicators will display eights (8's) in each digit position. All light-operated switches on the Operator Control Panel will remain lighted during this test.

Normally this test is performed with the MODE SELECT key in the TEACH position.

—CAUTION—

Pressing the LAMP TEST during operating modes other than TEACH will cause the hydraulics to shut down and the operating program to stop.

Cassette & HHTU Connectors. The HHTU is used primarily to record and modify positions for precise control from a remote station. The HHTU utilizes a thumb pressure contact switch called a "dead man switch" as a safety measure for operating the robot arm while standing in the vicinity of its' working area. In the event unwanted motions are encountered which may cause damage or personal injury, releasing the "dead man switch" will cause the hydraulics to automatically shut down and immediately stop the arm motion. (HHTU = HAND HELD TEACHING UNIT)

When the HHTU is connected to the Program Teach Panel all manual controls are automatically transferred to the HHTU. Two procedures are outlined which describe the method for recording new positions and for modifying existing positions with the use of the HHTU.

<u>PROCEDURE:</u> (Recording Positions)
(a) Mode Select key to the TEACH position.

(b) Select the desired Program Number to be used.

(c) Establish and <u>record</u> the HOME position (position 1).

Manual Controls

The Manual Controls are used for setting, recording, and modifying the robot arm positions. These controls consist of axis request, axis direction, record position, modify position, and next position number controls.

Setting the Position. Each one of the seven (7) possible axes has its specific request control. All axes are identified with yellow lighted pushbuttons for operator selection.

(a) <u>Axis Request:</u> When a change in a particular axis is desired, the corresponding Axis Request pushbutton should be selected. Once pressed the Axis Request control will light as an indication of the request.

NOTE: If an incorrect axis has been selected, pressing the desired Axis Request control will automatically change to the intended axis.

(b) <u>Axis Direction:</u> The Axis Direction controls are identified as ADVANCE and RETRACT. These controls will move the robot arm in the corresponding direction relative to the particular Axis Requested. Depressing and <u>holding</u> the Axis Request control will result in acceleration to predetermined velocity, while tapping the control will move the axis one count for each tap of the control.

NOTE: Motion of the robot arm requires hydraulic assisted devices and therefore the HYDRAULIC UNLOCK and PUMP ON Control on the Operator Control Panel must be actuated.

When moving the robot arm with the Advance or Retract Controls, either control must be <u>pressed</u> and <u>held</u> until the robot arm arrives at the desired position. If the direction of motion is opposite the intended direction, operating the other Axis Direction control will provide the desired motion.

Movement of the robot arm to the desired position is a result of operating the individual Axis Request and Axis Direction Controls until their combined motions place the robot arm in the desired position, that position should then be <u>recorded</u>.

Recording the Position. Motion of the robot arm can be achieved by operating the manual axis request and axis direction controls in any of the MODE SELECT positions, other than the Single Step Position. However, to record a position the MODE SELECT KEY should be placed in the TEACH position.

SAMPLE 1: Assume the robot arm is in the position A as shown by the heavy solid lines of the drawing in Fig. 5-4. The objective is to record the two positions indicated by the "x".

PROCEDURE:

(a) Place the MODE SELECT key in the TEACH position.
(b) Activate the hydraulics.
(c) Select the desired program number which is to be used.
(d) Select the HORZ Axis Request and operate the AD-VANCE CONTROL until the arm is in the desired outward position.
(e) Select the SWING Axis Request & operate the AD-VANCE Control until the arm is over the target. (1)

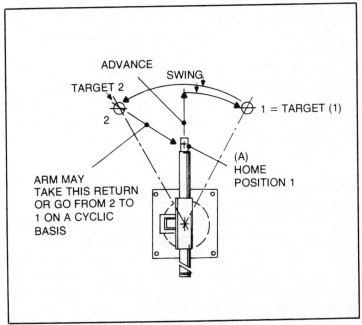

Fig. 5-4. Manual movement of the Prab Versatran robot arm (courtesy Prab Conveyors).

(f) Press the RECORD POS control to record the position in memory (position indicator will display the position number just recorded.)

(g) Select the Swing Axis Request and operate the RETRACT Control (if necessary) until the arm is directly over the second target.

(h) Press the RECORD POS Control to record the position in memory.

NOTE: Always attempt to record position number 1 (A) as the HOME POSITION.

Next Position Number Request. The Next Position Number control is used to select any particular recorded position within the program. Requesting a Next Position Number can only be done in the TEACH mode of operation.

PROCEDURE:

(a) Place the MODE SELECT key in the TEACH position.

(b) Select the desired program number which is to be used.

(c) Press the NEXT POSITION NUMBER control (the lamp will light as an indication of the selection).

(d) Select the desired position number on the keyboard. The Position Indicator will display the position number requested and the JOG lamp will light.

Jog to Position. After making the Next Position number request, it may be desirable to move the robot arm to the position requested. A simple procedure for testing the movement of the robot arm between any two positions is provided below.

— CAUTION —

Keep in mind, the Versatran System is a point-to-point system. Consequently, the combined motion of the robot arm will take the most direct route from one position to the next. Care must be taken to assure that the robot arm has a free path to travel. A JOG to Position, after making a position number request, will test the path of travel for the robot arm.

(a) Place the MODE SELECT key in the TEACH position.

(b) Select the desired program number to be used.

(c) Select the Next Position Number of interest.

(d) Press and <u>hold</u> the JOG switch button until the arm arrives in position. (When the robot arm arrives in position the JOG lamp will go off).

(e) Select the Next Position Number of interest and repeat step (d).

NOTE: If any obstructions are within contact of the robot arm, releasing the JOG pushbutton will stop the motion of the arm. Travel between these two (2) positions cannot be accomplished and therefore the robot arm will have to be programmed to bypass the obstacle. (See Fig. 5-5)

Modifying the Position. After setting and recording a position in memory it may be learned that the particular position is not quite correct. The position can quickly and simply be modified without the need to change the program.

PROCEDURE:

(a) Place the MODE SELECT key in the TEACH position.

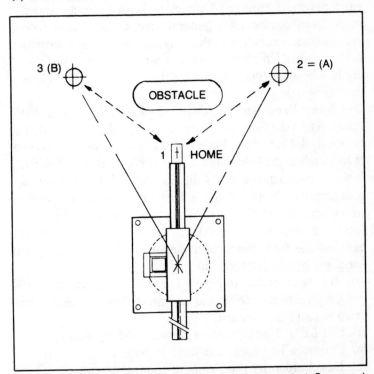

Fig. 5-5. Programmed arm moves around obstacle (courtesy Prab Conveyors). The robot arm comes to an abrupt stop at the home position and immediately resumes its motion toward the next position (3) or (2).

(b) Select the desired program number to be used.

(c) Select the Next Position Number of interest.

(d) Operate the Axis Request and Axis Direction controls until the robot arm is in the desired position.

(e) Press the MODIFY POS Control. The Control Unit will accept the modified position in memory for the position number displayed on the indicator.

NOTE: If attempt is made to modify the position with the RECORD POS control, the requested position will not be modified. The Record Position Control will record only new positions.

Function Controls

The Function Controls are used for establishing and entering a set of coded instructions for the Control Unit to execute in performing a particular task. An instruction code is made from combining a general function command with a specific function value. In this way, an instruction is provided to the Control Unit which directs a defined action. The following is a brief introduction to instruction code sets including examples using the instruction code.

Next Program Number: (Instruction Code NPN ..xx). When the Function Command NEXT PROG NUMBER is selected it is to be followed with a function value of no less than 1 and no greater than 64. The NEXT PROG NUMBER function command requests the Control Unit to go to another program in memory, while the function value (1-64) specifies which program in memory. When the Control Unit encounters an instruction code NPN ..XX, the Control Unit will execute the first instruction (sequence 1) of the program specified by the function value.

If it is desirable to repeat the same program, the NPN ..XX instruction code should contain the number of the program making the request.

EXAMPLE 1. Next Program Number (NPN ..XX)

(a) Purpose: To repeat the same program.

(b) Detail: Identify the Program Number to run.

STEP 1: REPEAT THE PROCESS

(c) Source Program:

Sequence No.	Function Command	Function Value	Enter	Instruction Code
BEGIN;	Prog Number	1	:	(.;PGN. . .1:)
1;	Next Prog Number	1	:	(1:NPN. . .1:)

NOTE: A POSITION MUST BE RECORDED BEFORE ANY PROGRAM CAN BE VALID.

NOTE: Although the source program of example 1 is a single instruction program, if "taught" to the Control Unit, it would be a perfectly acceptable program. The instruction to do program 1 would be repeated indefinitely until the program was stopped.

(d) Teach:
—Mode Select key to the Teach position.
—ENTER the Source Program in sequence starting with BEGIN.
—Record a Position (Press RECORD).

(e) Check:
—Mode SELECT key to the SINGLE STEP position.
—Press the SINGLE STEP FWD Control to review the instruction.

Call Position (Instruction Code CPN XXXX). When the Function Command CALL POS is selected, it is to be followed with a function value no less than 1 and no greater than 9999. The CALL POS function command requests the Control Unit to direct the movement of the robot arm, while the function value (1—9999) specifies the particular position.

When the Control Unit encounters an instruction code CPN XXXX it will cause the robot arm to move in the most direct route to the position identified.

EXAMPLE 2. (Call Position CPN XXXX)

(a) Purpose: To move the robot arm from one position to another and repeat the process.

(b) Detail: Identify the program number to be run.
STEP 1: START AT HOME AND GO TO NEXT POSITION.
STEP 2: RETURN HOME
STEP 3: REPEAT THE PROCESS

NOTE: It is advisable to establish position 1 as the Home position. Position 1 is always the beginning position of any program. Also, it is good practice to place the Home Position (position 1) at a point which is out of the way. When the last position called in the program is the Home Position the robot arm will come to rest (out of the way) when a COMPLETE PROGRAM THEN STOP is activated.

(c) Source Program:

Sequence No.	Function Command	Function Value	Enter	Instruction Code
BEGIN;	Prog Number	2	:	(.;PGN. . .2:)
1;	Call Position	2	:	(1;CPN. . .2;)
2;	Call Position	1	:	(2;CPN. . .1:)
3;	Next Prog Number	2	:	(3;NPN. . .2:)

(d) Teach:
   (i) Mode Select key to the TEACH position.
   (ii) ENTER the Source Program in sequence starting with BEGIN.
   (iii) SET and RECORD positions 1 and 2.

(e) Check:
   (i) Mode Select key to the SINGLE STEP position.
   (ii) Press the SINGLE STEP FWD Control to review each instruction in sequence.

NOTE: If desired, the JOG switch button can be operated each time it lights to move the robot arm to each new position. This practice is advisable to assure the operator that the arm has a free path to travel.

(f) Operate the Program:
   (i) Mode Select key to the AUTO RUN position.
   (ii) Press JOG to move robot arm to HOME (Position 1).
   (iii) Press RUN PROGRAM to execute the program.

NOTE: The program in example 2 will operate the Control Unit's predetermined speed. Faster or Slower operation will require instructing the Control Unit Velocity and Acceleration Codes.

     Set Velocity: (Instruction Code SVL. .XX). When the Function Command SET VEL is to be followed with a functional value no less than 1 and no greater than 16. The SET VEL function command requests the Control Unit to direct the motion of the robot arm at a speed which is specified by the function value (1-16). The function values

1-16 are incremented from ½ IPS to 37½ IPS with function value 1 representing ½ IPS and function value 16 representing 37½ IPS. See Table (5-1) for a complete listing of velocity selections. (IPS = inches/per/second)

<u>NOTE</u>: Once a velocity value is selected, it remains in effect for all following movements until such time as a new value is selected. This is also true when the Control Unit transfers to another program in memory.

Generally, when specifying a velocity for the arm to move, an acceleration value should also be indicated.

Set Acceleration/Deceleration: (Instruction Code SAD. .XX). When the Function Command SET ACC/DEC is selected, it is to be followed with a functional value no less than 1 and no greater than 16. The SET ACC/DEC function command requests the Control Unit to direct the rate of change of speed of the robot arm to its maximum specified velocity. The acceleration/deceleration rates are specified by the function values (1-16) following the function command.

The function values (1-16) are incremented from 15 inches/sec/sec/ to 600 inches/sec/sec/ with the value 1 representing 15 inches/sec/sec/ and the value of 16 representing 600 inches/sec/sec. See Table 5-1 for a complete listing of acceleration/deceleration selections.

It is common practice, and recommended, that the function values of velocity and acceleration are referred to with the velocity value first and then the ACC/DEC value last. (i.e., the values 5/4 represent velocity 5/acceleration 4). In this way, good communications with others involved with the program will be maintained.

The velocity and acceleration instructions (SVL. .XX and SAD. .XX) should generally precede the CALL position instruction (CPN XXXX) when movement to that position requires velocity and acceleration changes. When no VEL or ACC/DEC is specified, the Control Unit will resort to the present values of 5/4 as in the program of example 2.

<u>NOTE</u>: Once an ACC/DEC value is selected, it remains in effect for all following movements until such time as a new value is selected. This is also true when the Control Unit

Table 5-1. Velocity & ACC/DEC Values.

VELOCITY			FUNCTION VALUE	ACC/DEC		
LINEAR	SWING	WRIST		LINEAR	SWING	WRIST
37½"/SEC	170°/SEC	218°/SEC	16	600"/SEC/SEC	2700°/SEC/SEC	3500°/SEC/SEC
30"/SEC	135°/SEC	196°/SEC	15	540"/SEC/SEC	2430°/SEC/SEC	3150°/SEC/SEC
26"/SEC	118°/SEC	155°/SEC	14	480"/SEC/SEC	2160°/SEC/SEC	2800°/SEC/SEC
22½"/SEC	100°/SEC	132°/SEC	13	420"/SEC/SEC	1890°/SEC/SEC	2450°/SEC/SEC
19"/SEC	85°/SEC	110°/SEC	12	360"/SEC/SEC	1620°/SEC/SEC	2100°/SEC/SEC
15"/SEC	68°/SEC	88°/SEC	11	300"/SEC/SEC	1350°/SEC/SEC	1750°/SEC/SEC
12"/SEC	50°/SEC	65°/SEC	10	240"/SEC/SEC	1080°/SEC/SEC	1400°/SEC/SEC
8"/SEC	34°/SEC	44°/SEC	9	180"/SEC/SEC	810°/SEC/SEC	1050°/SEC/SEC
4"/SEC	18°/SEC	22°/SEC	8	120"/SEC/SEC	540°/SEC/SEC	700°/SEC/SEC
3½"/SEC	15°/SEC	19°/SEC	7	105"/SEC/SEC	472°/SEC/SEC	612°/SEC/SEC
3"/SEC	12°/SEC	16°/SEC	6	90"/SEC/SEC	405°/SEC/SEC	525°/SEC/SEC
2½"/SEC	10°/SEC	14°/SEC	5	75"/SEC/SEC	338°/SEC/SEC	438°/SEC/SEC
2"/SEC	8°/SEC	11°/SEC	4	60"/SEC/SEC	270°/SEC/SEC	350°/SEC/SEC
1½"/SEC	6°/SEC	8°/SEC	3	45"/SEC/SEC	200°/SEC/SEC	260°/SEC/SEC
1"/SEC	4°/SEC	5°/SEC	2	30"/SEC/SEC	135°/SEC/SEC	175°/SEC/SEC
½"/SEC	2°/SEC	3°/SEC	1	15"/SEC/SEC	68°/SEC/SEC	88°/SEC/SEC

*The velocity and acc/dec values of Table are equivalent to a single axis motion. Multiple axis motions proportionally reduce the velocity and acc/dec values.

164

transfers to another program in memory.

EXAMPLE 3: Set Velocity & Set ACC/DEC (SVL. .XX & SAD. .XX)

(a) Purpose: To move the robot arm from Home position to another position and return home at a medium high velocity and acceleration.

(b) Detail:

Identify the Program Number to be used.

STEP 1: MOVE AT MED/HI VELOCITY

STEP 2: MOVE WITH MED/HI ACC/DEC

STEP 3: START AT HOME AND MOVE TO NEXT POSITION.

STEP 4: RETURN TO HOME

STEP 5: REPEAT THE PROCESS

(c) Source Program:

Sequence No.	Function Command	Value	Enter	Instruction Code
BEGIN;	Prog Number	3	:	(.;PGN. . .3:)
1;	SET VEL	12	:	(1;SVL. .12:)
2;	SET ACC /DEC	10	:	(2;SAD. .10:)
3;	CALL POS	2	:	(3;CPN. . .2:)
4;	CALL POS	1	:	(4;CPN. . .1:)
5;	Next Prog Number	3	:	(5;NPN. . .3:)

(d) Teach:

(i) Mode Select key to the TEACH position.

(ii) ENTER the Source Program in sequence starting with BEGIN.

(iii) SET and RECORD positions 1 and 2.

(e) Check:

(i) Mode Select key to the SINGLE STEP position.

(ii) Press the SINGLE STEP FWD Control to review each program instruction in sequence. If desired, the JOG switch button can be operated each time it lights to move the robot arm to each new position. This practice is advisable to assure the operator that the robot arm has a free path to travel.

(f) Operate the Program:

(i) Mode Select key to the AUTO RUN position.

(ii) Press JOG to move the arm to HOME (position 1).

(iii) Press RUN PROGRAM to execute the program.

EXAMPLE 3: Set Velocity & Set ACC/DEC (SVL. .XX & SAD. .XX)

(a) Purpose: To move the robot arm from Home position to the next position at a medium-low velocity and return Home at a high velocity, then repeat the process.

(b) Detail:

Identify the Program Number to be run.

STEP 1: MOVE AT MED/LOW VELOCITY

STEP 2: MOVE WITH MED/LOW ACC/DEC

STEP 3: START FROM HOME AND GO TO NEXT POSITION

STEP 4: MOVE AT HIGH VELOCITY

STEP 5: MOVE WITH HIGH ACC/DEC

STEP 6: GO HOME

STEP 7: REPEAT THE PROCESS

(c) Source Program:

Sequence No.	Function Command	Function Value	Enter	Instruction Code
BEGIN;	Prog Number	31	:	(.;PGN. . .31:)
1;	SET VEL	6	:	(1;SVL. . .6:)
2;	SET ACC /DEC	6	:	(2;SAD. . .6:)
3;	CALL POS	2	:	(3;CPN. . .2:)
4;	SET VEL	14	:	(4;SVL. .14:)
5;	SET ACC /DEC	12	:	(5;SAD. .12:)
6;	CALL POS	1	:	(6;CPN. . .1:)
7;	Next Prog Number	31	:	(7;NPN. .31:)

(d) Teach:

(i) Mode Select key to the TEACH position.

(ii) ENTER the Source Program in sequence starting with BEGIN.

(iii) SET and RECORD positions 1 and 2.

(e) Check:

(i) Mode Select key to the SINGLE STEP position

(ii) Press the SINGLE STEP FWD Control to review each instruction in sequence. If desired, the JOG switch button can be operated each time it lights to move the robot arm to each new position. This practice is advisable to assure the operator that the arm has a free path to travel.

(f) Operate the Program:

(i) Mode Select key to the AUTO RUN position.

(ii) Press JOG to move the Robot Arm to Home (position 1).

(iii) Press RUN PROGRAM to execute the instructions.

Round Off Position (Instruction Code ROP XXXX). When a function command ROUND OFF POSITION is selected, it is to be followed by a function value of no less than 1 and no greater than 9999. The ROUND OFF POS function command requests that the Control Unit direct the robot arm to the position specified by the function value. However, motion of the robot arm to and from the specified position is accomplished at program velocity. As the arm approaches the position, the deceleration value of the program is ignored, consequently, the robot arm arrives at the position still moving at maximum specified speed and abruptly stops. Further motion to another position ignores the acceleration value of the program and immediately assumes maximum specified speed.

SAMPLE 2 (see Fig. 5-5):

The desired motion is to move the robot arm between position 2 and 3. However, direct motions between these positions would cause the arm to strike the obstacle in its path. It is further desired to move the arm between position 2 and 3 with a constant velocity. Rounding off the Home position (1) and passing the robot arm through this position can essentially achieve the desired motion.

EXAMPLE 4: Round Off Position (ROP XXXX)

(a) Purpose: To move the robot arm between two (2) targets avoiding an obstacle in its path and repeat the process.

(b) Detail:

Identify the Program Number to be used.

STEP 1: START AT HOME AND GO TO TARGET A.

STEP 2: INITIALIZE VELOCITY (MED-LOW).

STEP 3: INITIALIZE RATE (MED-LOW).

STEP 4: CALL HOME.

STEP 5: GO TO TARGET B.

STEP 6: INITIALIZE VELOCITY (MED-HIGH).

STEP 7: INITIALIZE RATE (MED-HIGH).

STEP 8: CALL HOME.

STEP 9: REPEAT

167

<u>NOTE:</u> According to the detail of example 4, the robot arm will move from position 2 to 3 and continuously operate between these two positions. Movement from position 2 to position 3 is done with a med-low velocity and the return to position 2 is done with a med-high velocity. Position 1 is set as the home position such that when the robot arm stops at the end of the program, it is in a position that should be out of the way of the work area targets A and B).

Sequence No.	Function Command	Value	Enter	Instruction Code
BEGIN;	Prog Number	4	:	(.;PGN. . .4:)
1;	CALL POS	2	:	(1;CPN. . .2:)
2;	SET VEL	6	:	(2;SVL. . .6:)
3;	SET ACC /DEC	6	:	(3;SAD. . .6:)
4;	ROUND OFF POS	1	:	(4;ROP. . .1:)
5;	CALL POS	3	:	(5;CPN. . .3:)
6;	SET VEL	10	:	(6;SVL. .10:)
7;	SET ACC /DEC	9	:	(7;SAD. . .9:)
8;	ROUND OFF POS	1	:	(8;ROP. . .1:)
9;	Next Prog Number	4	:	(9;NPN. . .4:)

(d) <u>Teach:</u>
   (i) Mode Select key to the TEACH position.
   (ii) ENTER the Source Program starting with BEGIN.
   (iii) SET and RECORD position 1 (Home), 2 and 3.
(e) <u>Check:</u>
   (i) Mode Select key to the SINGLE STEP position.
   (ii) Press the SINGLE STEP FWD Control to review each instruction in sequence. If desired, the JOG switch button can be operated each time the lamp lights to move the robot arm to each new position. This practice is advisable to assure the operator that the arm has a free path to travel.
(f) <u>Operate the Program:</u>
   (i) Mode Select key to the AUTO RUN position.
   (ii) Press JOG to move the robot arm to Home (position 1).
   (iii) Press the RUN PROGRAM switch button to execute the program.

   Time Delay & Interlock Dwell Instructions (TDL NXXX & ILD. .XX; ILD.10X). When the TIME DELAY Function Command is selected, it is to be followed with a function value that identifies the timer number to be set at no less than .1 second and no greater than 99.9 seconds. The TIME DELAY requests one of ten timers to be set for a specific period of time and immediately begin counting down to zero. The period of time can be selected from 0.1 seconds to 99.9 seconds in 0.1 second intervals. Choosing one of ten timers is accomplished by recording the first digit (most significant digit) of the four digit number as the timer number and the remaining three digits with the desired time interval: i.e., timer number 3 to be set at 6.5 seconds; the function value should be set as 3065.

The timers are numbered from 0-9. When using timer number 0, it is not necessary to specify the timer number prior to setting the desired time delay value: i.e., timer number 0 desired to be set at 6.5 seconds; the function value should be set as . . 65. The INTLK DWELL function command has two significant purposes; (1) *Wait for external source command before proceeding*, and (2) *Wait for timer to time out before proceeding*. The WAIT FOR EXTERNAL SOURCE Command is determined by the function values 1-32 following the INTLK DWELL selection. This instruction requests that the Control Unit sit and wait until a signal is received from one of the selected 32 input lines. When the selected input line becomes "true", the process will continue, until such time the program will be stalled; i.e., wait for part to get into position (assume position signal is received on line one); the function value should be set as . . . 1; when the part arrives in position, a signal will be sent on line one which informs the Control Unit to resume the program.

The WAIT FOR TIME OUT Command is determined by the function value 10 plus the particular timer number. This instruction requests the the Control Unit stop the program until the specified timer reaches zero: i.e., wait for timer number 3 to time to zero; the value should be set as 103; function when timer number 3 reaches zero, the Control Unit will resume the program.

EXAMPLE 5: Time Delay & Interlock Dwell (TDL NXXX & ILD.XXX) Wait for Time Out (ILD.10X)

(a) <u>Purpose:</u> To move the robot arm to position 2 and wait for two (2) seconds and then move to position 3 and wait for two (2) seconds. REPEAT THE PROCESS.

(b) <u>Detail:</u>

Identify the program number to be used.

STEP 1: START AT HOME AND MOVE TO TARGET A.

STEP 2: SET TIME (2 SECONDS)

STEP 3: WAIT TIME

STEP 4: INITIALIZE VELOCITY (MED-LOW)

STEP 5: INITIALIZE RATE (MED-LOW)

STEP 6: GO HOME

STEP 7: GO TO TARGET B.

STEP 8: SET TIME (2 SECONDS)

STEP 9: WAIT TIME

STEP 10: INITIALIZE VELOCITY (MED-HIGH)

STEP 11: INITIALIZE RATE (MED-HIGH)

STEP 12: GO HOME

STEP 13: REPEAT THE PROCESS

(c) <u>Source Program:</u>

Sequence No.	Function Command	Value	Enter	Instruction Code
BEGIN;	Prog Number	5	:	(.;PGN. . .5:)
1;	CALL POS	2	:	(1;CPN. . .2:)
2;	TIME DELAY	20	:	(2;TDL. .20:)
3;	INTLK DWELL	100	:	(3;ILD;100:)
4;	SET VEL	6	:	(4;SVL. . .6:)
5;	SET ACC /DEC	6	:	(5;SAD . . .6:)
6;	CALL POS	1	:	(6;CPN. . .1:)
7;	CALL POS	3	:	(7;CPN. . .3:)
8;	TIME DELAY	20	:	(8;TDL. .20:)
9;	INTLK DWELL	100	:	(9;ILD.100:)
10;	SET VEL	10	:	(10;SVL. .10:)
11;	SET ACC /DEC	9	:	(11;SAD . . .9:)
12;	CALL POS	1	:	(12;CPN. . .1:)
13;	Next Prog Number	5	:	(13;NPN. . .5:)

(d) <u>Teach:</u>

(i) Mode Select key to the TEACH position.

(ii) ENTER the Source Program starting with BEGIN.

(iii) SET and RECORD positions 1 (HOME), 2 and 3.

(e) <u>Check:</u>

(i) Mode Select key to the SINGLE STEP position

(ii) Press the SINGLE STEP FWD to review each instruction in sequence. If desired, press the JOG pushbutton each time the lamp lights to move the robot arm to its new position. This practice is advisable to assure the operator that the arm has a free path to travel.

(f) Operate the Program:

   (i) Mode Select key in the AUTO RUN position.

   (ii) Press JOG to move the Robot Arm to Home, (position 1), Fig. 5-5.

   (iii) Press the RUN PROGRAM switch button to execute the program.

EXAMPLE 5.1: Interlock Dwell (Wait For External Source)

(a) Purpose: To move the robot arm from target A through HOME to target B. Wait at target B until ready to accept payload and repeat the process.

NOTE: The program purpose would be to release the payload at the robot arm tooling end only when the receptacle is "ready" to receive the payload.

(b) Detail (refer back to Fig. 5-5):

Identify the program number to be used.

STEP 1: START AT HOME AND GO TO POSITION 2.

STEP 2: INITIALIZE VELOCITY.

STEP 3: INITIALIZE RATE.

STEP 4: GO HOME.

STEP 5: GO TO POSITION 3.

STEP 6: WAIT FOR "READY TO RECEIVE PAYLOAD".

STEP 7: GO HOME.

STEP 8: REPEAT THE PROCESS.

(c) Source Program.

Sequence No.	Function Command	Value	Enter	Instruction Code
BEGIN;	Prog Number	51	:	(.;PGN. .51:)
1;	CALL POS	2	:	(1;CPN. . .2:)
2;	SET VEL	6	:	(2;SVL. . .6:)
3;	SET ACC /DEC	6	:	(3;SAD . . .6:)
4;	CALL POS	1	:	(4;CPN. . .1:)
5;	CALL POS	3	:	(5;CPN. . .3:)
6;	INTLK DWELL	1	:	(6;ILD. . .1:)
7;	CALL POS	1	:	(7;CPN. . .1:)
8;	Next Prog Number	51	:	(8;NPN. .51:)

NOTE: TEACHING the Source Program of example 5.1 to the Control Unit without a valid external command signal on line 1 will result in the robot arm motion to cease at position 3 where it will wait indefinitely for the resume command. To execute the above program without a valid external command signal; press CONTROLLED STOP, then RUN PROGRAM Controls on the Operator Panel. The interlock will be bypassed and the program will resume with the next program sequence instruction.

Output Command On & Output Command Off Instructions: (OCON. .XX & OCOF. .XX). When either an OUTPUT COMMAND ON or OFF is selected, it is to be followed by a function value no less than 1 and no greater than 32.

The OUTPUT CMD ON or OUTPUT CMD OFF function command requests that the Control Unit energizes or deenergizes the mechanism connected to the output line specified by the function value. Typically this would be the end-of-arm tooling designed to perform a specific task (such as grippers, welders, drills, etc.). Also the specified output line can be various other equipment (such as milling machines, motors, etc., or external devices which are to be controlled with the process to performing the task).

Other functions of the OUTPUT CMD ON/OFF are identified and described in the following section of this manual.

EXAMPLE 6: Output Command On and Output Command Off
(Instruction Codes: OCON. .XX & OCOF . .XX)

(a) Purpose: To move the robot arm to target A and pick-up part. Move part to target B avoiding obstacle and release part. Repeat the process (again refer to Fig. 5-5).

(b) Detail:
Identify the program number to be used.
STEP 1: START AT HOME AND MOVE OVER TARGET.
STEP 2: SET TIME ½ SECOND
STEP 3: WAIT TIME

STEP 4: PICK UP PART
STEP 5: SET TIME ½ SECOND
STEP 6: WAIT TIME
STEP 7: INITIALIZE VELOCITY (MED-LOW)
STEP 8: INITIALIZE RATE (MED-LOW)
STEP 9: GO HOME
STEP 10: GO TO TARGET B
STEP 11: SET TIME ½ SECOND
STEP 12: WAIT TIME
STEP 13: RELEASE PART
STEP 14: SET TIME ½ SECOND
STEP 15: WAIT TIME
STEP 16: INITIALIZE VELOCITY (MED-HIGH)
STEP 17: INITIALIZE RATE (MED-HIGH)
STEP 18: GO HOME
STEP 19: REPEAT THE PROCESS

(c) Source Program:

Sequence No.	Function Command	Value	Enter	Instruction Code
BEGIN;	Prog Number	6	:	(.;PGN. . .6:)
1;	CALL POS	2	:	(1;CPN. . .2:)
2;	TIME DELAY	5	:	(2;TDL. . .5:)
3;	INTLK DWELL	100	:	(3;ILD.100:)
4;	OUTPUT COMD ON	1	:	(4;OCON. . .1:)
5;	TIME DELAY	5	:	(5;TDL. . .5:)
6;	INTLK DWELL	100	:	(6;ILD.100:)
7;	SET VEL	6	:	(7;SVL. . .6:)
8;	SET ACC /DEC	6	:	(8;SAD . . .6:)
9;	CALL POS	1	:	(9;CPN. . .1:)
10;	CALL POS	3	:	(10;CPN. . .3:)
11;	TIME DELAY	5	:	(11;TDL. . .5:)
12;	INTLK DWELL	100	:	(12;ILD.100:)
13;	OUTPUT CMD OFF	1	:	(13;OCOF. . .1:)
14;	TIME DELAY	5	:	(14;TDL. . .5:)
15;	INTLK DWELL	100	:	(15;ILD;100:)
16;	SET VEL	10	:	(16;SVL. .10:)
17;	SET ACC /DEC	9	:	(17;SAD . . .9:)
18;	CALL POS	1	:	(18;CPN. . .1:)
19;	Next Prog Number	6	:	(19;NPN. . .6:)

(d) Teach:
(i) Mode Select key to the TEACH position.
(ii) ENTER the Source Program starting with BEGIN.
(iii) SET and RECORD positions 1 (Home), 2 and 3.

(e) Check:
(i) Mode Select key to the SINGLE STEP position.
(ii) Press the SINGLE STEP FWD Control to review each instruction in sequence. If desired, press JOG switch button each time the lamp lights to move the robot arm to its new position. This practice is advisable to assure the operator that the arm has a free path to travel.

(f) Operate the Program:
(i) Mode Select key to the AUTO RUN position.
(ii) Press JOG to move the ROBOT Arm to HOME (position 1).
(iii) Press the RUN PROGRAM switch button to execute the program.

Call Subroutine Instruction (CSR. .XX). When a Function Command CALL SUBROUTINE is selected, it is to be followed with a function value no less than one and no greater than 64. The CALL SUBROUTINE function command requests a departure from the operating program to another program in memory. Any departure will require the eventual return to the main program. A return to the main program is accomplished when a NEXT PROGRAM NUMBER instruction is encountered which has its function value the same as the program making the request (NPN SAME:). When the return to the main program is accomplished, the main program will resume with the next instruction following the CALL SUBROUTINE instruction. A single subroutine can be repeated indefinitely as long as a return to the main program occurs before calling the subroutine again.

A return to the calling program from any subroutine is executed when the subroutine encounters a NEXT PROG NUMBER (NPN) instruction which has the program number specified to be the same as the program making the request. The executive program logic executes a sequence of instructions which accomplish the task as shown below when any NPN instruction is encountered.

All subroutines return only to their calling program at the program sequence number following its call subroutine instruction.

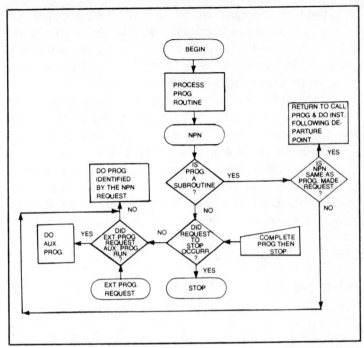

Fig. 5-6. A Flowchart for the Versatran robot Executive Program (courtesy Prab Versatran).

One subroutine calling from another subroutine is referred to as "nesting". Nested subroutines form a chain of subroutines which are linked to the original calling program. If the second subroutine calls for another subroutine prior to returning to the first subroutine, the chain, or "nesting", grows. Up to twenty (20) levels of subroutines can be nested in this manner.

The return to the original calling program, from where the first subroutine was called, is accomplished in the reverse order in which they were called.

NESTED SUBROUTINES:

MAIN PROGRAM #1	SUBROUTINE #2	SUBROUTINE #3
BEGIN ; PGN. . .1:	BEGIN ; PGN. . .2:	BEGIN ; PGN. . .3:
1 ; TDL. .10:	1 ; TDL. .20:	1 ; TDL. . .5;
2 ; ILD. 100:	2 ; ILD. 100:	2 ; ILD. 100:
3 ; CSR. . .2:	3 ; CSR. . .3:	3 ; NPN. . .3:
4 ; NPN. . .1:	4 ; NPN. . .2:	

The main program calls for subroutine #2 at instruction sequence 3. Similarly, subroutine #2 calls for subroutine #3 at its third instruction sequence prior to returning to the main program. Therefore, subroutine #3 is nested to the main program through subroutine #2.

Once subroutine #3 has completed its program, it returns to subroutine #2 at instructions sequence 4. Consequently, subroutine #2 returns to the main program at instruction sequence 4 and the process is repeated.

Twenty (20) levels can be nested in this manner prior to any one subroutine returning to its calling program. Subroutines which are not nested can be repeated indefinitely, as long as they return to their calling program prior to being executed again.

UNNESTED SUBROUTINES:

MAIN PROGRAM	SUBROUTINE #2	SUBROUTINE #3
BEGIN ; PGN. . .1:	BEGIN ; PGN. . .2:	BEGIN ; PGN. . .3:
1 ; TDL .10:	1 ; TDL. .20:	1 ; TDL. . .5:
2 ; ILD .100;	2 ; ILD .100:	2 ; ILD .100:
3 ; CSR. . .2:	3 ; NPN. . .2:	3 ; NPN. . .3:
4 ; CSR. . .3:		
5 ; NPN. . .1:		

The main program calls for subroutine #2 at instruction sequence 3. Once subroutine #2 completes its program, it returns to the main program at instruction sequence 4. At that time, subroutine #3 is called (from the main program) and executed. When its program is finished, subroutine #3 returns directly to the main program at instruction sequence 5.

Once a CALL SUBROUTINE instruction is encountered, all further movement of the robot arm is relative to the position of the arm at the time of the instruction: i.e., the Control Unit considers the position of the arm where the departure occurred as the HOME POSITION (position 1) of the subroutine.

The samples as shown in Fig. 5-7a and 5-7b are showing the motions of the robot arm for the main program and the subroutine program respectively.

If a CALL SUBROUTINE instruction occurs while the robot arm is at position 3 of the main program, that position

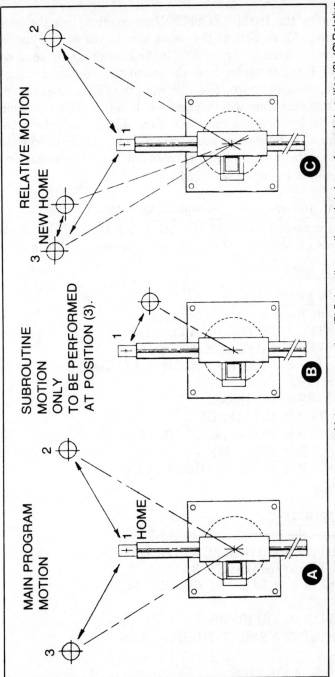

Fig. 5-7. Relative motions of Versatran robot arm. (A) Main program motion. (B) Subroutine motion only to be performed at position (3). (C) Relative motion (courtesy PRAB Versatran).

becomes the HOME POSITION or position 1 of the subroutine. All motion of the robot arm in the subroutine is therefore relative to position 3 in the main program (see Fig. 5-7c). Upon returning from the subroutine to the main program, the next instruction following the CSR instruction of the main program will be executed. It is therefore advisable that the last position called for in the subroutine is the same position as the starting position of the subroutines in order to resume the program with the robot arm in the position it was when the departure occurred.

EXAMPLE 7: Call Subroutine (CSR. .XX)

(a) Purpose: To OPERATE THE MAIN PROGRAM with a CALL SUBROUTINE instruction occurring at position 3.

(b) Detail:

MAIN PROGRAM
Identify the program number used.
    STEP 1: INITIALIZE VELOCITY
    STEP 2: INITIALIZE RATE
    STEP 3: START AT HOME & GO TO POS 2
    STEP 4: GO HOME
    STEP 5: GO TO POS 3
    STEP 6: DO SUBROUTINE 12
    STEP 7: GO HOME
    STEP 8: REPEAT THE PROCESS

SUBROUTINE
Identify the program number used.
    STEP 1: INITIALIZE VELOCITY
    STEP 2: INITIALIZE RATE
    STEP 3: START AT HOME & GO TO POS 2
    STEP 4: GO HOME
    STEP 5: REPEAT THE PROCESS

(c) Source Program:

MAIN PROGRAM

Sequence No.	Function Command /	Value	Enter	Instruction Code
BEGIN;	Prog Number	7	:	(.;PGN. . .7:)
1;	SET VEL	6	:	(1;SVL. . .6:)
2;	SET ACC /DEC	6	:	(2;SA/D. . .6:)
3;	CALL POS	2	:	(3;CPN. . .2:)
4;	CALL POS	1	:	(4;CPN. . .1:)
5;	CALL POS	3	:	(5;CPN. . .3:)
6;	CALL SUBROUTINE	12	:	(6;CSR. . .12:)
7;	CALL POS	1	:	(7;CPN. . .1:)
8;	Next Prog Number	7	:	(8;NPN. . .7:)

SUBROUTINE

Sequence No.	Function Command /	Value	Enter	Instruction Code
BEGIN;	Prog Number	12	:	(.;PGN. .12:)
1;	SET VEL	10	:	(1;SVL. .10:)
2;	SET ACC /DEC	10	:	(2;SA/D. .10:)
3;	CALL POS	2	:	(3;CPN. . .2:)
4;	CALL POS	1	:	(4;CPN. . .1:)
5;	Next Prog Number	12	:	(5;NPN. .12:)

(d) <u>Teach:</u>

(i) Mode Select key to the TEACH position.

(ii) ENTER both Source Program starting with BEGIN.

(iii) SET and RECORD positions 1 (HOME), and 2 on the subroutine program.

(e) <u>Check:</u>

(i) Mode Select key to the SINGLE STEP position.

(ii) Press the SINGLE STEP FWD Control to review each instruction in sequence.

CAUTION: It is recommended that the JOG switch pushbutton is operated each time it lights since the subroutine arm motions are RELATIVE TO THE POINT OF DEPARTURE. In this way avoiding obstructions or pegging the arm against a mechanical stop can be assured.

(f) <u>Operate the Program:</u>

(i) Mode Select key to the AUTO RUN position.

(ii) Press JOG to move the Robot Arm HOME (position 1).

(iii) Press the RUN PROGRAM switch button to execute the program.

Skip Sequence Number Instruction (SKIP. .XX).
The SKIP SEQUENCE NUMBER instruction is a conditional

jump which bypasses the next instruction in the program sequence when the condition of the specified input line is "true."

When the SKIP Function Command is encountered, the Control Unit will check the condition of the line specified by the function value (1-32). If the line is "true," the next instruction in the programming sequence is skipped, consequently, the one following the skipped instruction will be executed. However, if the specified input line is "not true" the next instruction will not be skipped.

EXAMPLE 8: Skip Sequence Number (Instruction Code SKIP. .XX)

(a) Purpose: To move the robot arm from target A to target B avoiding any obstacle and check target B for a part. If part is in position pick-up and move to target A; if part is not in position do subroutine UNTIL PART IS IN POSITION, THEN REPEAT THE PROCESS.

(b) Detail:
   Identify the program number to be used.

	STEP	1:	INITIALIZE VELOCITY
	STEP	2:	INITIALIZE RATE
	STEP	3:	START AT HOME AND GO TO TARGET A
	STEP	4:	RELEASE THE PART
	STEP	5:	GO HOME
	STEP	6:	GO TO TARGET B
	STEP	7:	CHECK FOR PART; : YES
BACK:	STEP	8:	DO SUBROUTINE
	STEP	9:	CHECK FOR PART: YES 1
YES:	STEP	10:	GO BACK TO 8 : BACK
	STEP	11:	GO TO TARGET B
	STEP	12:	PICK UP PART
	STEP	13:	GO HOME
	STEP	14:	REPEAT THE PROCESS

NOTE: Upon return from a subroutine, the next instruction following the CALL SUBROUTINE is executed. The program desires to repeat the subroutine if the part is not at target B. Therefore after returning from the subroutine call the next instruction is to check for the part again, if it is not there an immediate jump back to step 8 (Do subroutine) instruction is executed. An immediate skip instruction is coded by a Function Command SKIP SEQ NUMBER and a function value of 1000 plus the program sequence number desired. (Hence STEP 10 of the source program, SKIP 1008:)

(c) Source Program:

Sequence No.	Function Command	Value	Enter	Instruction Code
BEGIN	PROG NUMBER	8	:	(.:PGN. . .8:)
1	SET VEL	6	:	(1;SVL. . .6:)
2	SET ACC /DEC	6	:	(2;SAD . . .6:)
3	CALL POS	2	:	(3;CPN. . .2:)
4	OUTPUT CMD OFF	1	:	(4;OCOF. . .1:)
5	CALL POS	1	:	(5;CPN. . .1:)
6	CALL POS	3	:	(6;CPN. . .3:)
7	SKIP SEQ NUMBER	2	:	(7;SKIP. . .2:)
8	CALL SUBROUTINE	12	:	(8;CSR. .12:)
9	SKIP SEQ NUMBER	2	:	(9;SKIP. . .2:)
10	SKIP SEQ NUMBER	1008	:	(10;SKIP 1008:)
11	CALL POS	3	:	(11;CPN. . .3:)
12	OUTPUT CMD ON	1	:	(12;OCON. . .1:)
13	CALL POS	1	:	(13;CPN. . .1:)
14	NEXT PROG NUMBER	8	:	(14;NPN. . .8:)

NOTE: The Source Program of example 8 assumes that the part pickup device is connected to output line 1 and the part position device is connected to input line 2.

NOTE: Program sequence number 11 recalls position 3 to be sure that returning from the subroutine places the arm over the part pickup target.

EXAMPLE 9: Program To Use All Instructions Used In Examples 1 Through 8

(a) Purpose: To get part at target A and move to target B when part is "ready". If part is not ready, go to target C and spray target area with short burst. Check for part ready, if not continue spray.

Note: The program of example 9 assumes:
Part pick-up Device on line 1.
Part Position Sensor on line 2.
Spray gun actuator in line 3.

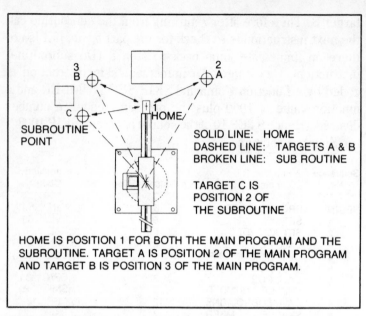

SUBROUTINE
POINT

SOLID LINE: HOME
DASHED LINE: TARGETS A & B
BROKEN LINE: SUB ROUTINE

TARGET C IS
POSITION 2 OF
THE SUBROUTINE

HOME IS POSITION 1 FOR BOTH THE MAIN PROGRAM AND THE
SUBROUTINE. TARGET A IS POSITION 2 OF THE MAIN PROGRAM
AND TARGET B IS POSITION 3 OF THE MAIN PROGRAM.

Fig. 5-8. Versatran robot's arm motion when waiting for a part (courtesy PRAB
Versatran).

INSTRUCTION	DETAIL

Main Program 9

	INSTRUCTION	DETAIL
	BEGIN ; PGN 9	ESTABLISH PROGRAM NUMBER
	1 ; SVL 6:	INITIALIZE VELOCITY (MED-LOW)
	2 ; SAD . . . 6:	INITIALIZE RATE (MED-LOW)
	3 ; SKIP. . . . 2: YES	CHECK FOR PART
	4 ; SRC 14:	DO PROGRAM 14
YES	5 ; CPN 2:	GO OVER PART
	6 ; TDL 5:	SET TIME
	7 ; ILD . . .100:	WAIT TIME
	8 ; CPN 3:	GO TO PART
	9 ; OCON. . . 1:	RETRIEVE PART
	10 ; TDL 5:	SET TIME
	11 ; ILD . . .100:	WAIT TIME
	12 ; CPN 2:	LIFT PART
	13 ; TDL 5:	SET TIME
	14 ; ILD . . .100:	WAIT TIME
	15 ; CPN 1:	GO HOME
	16 ; CPN 4:	GO OVER DUMP
	17 ; TDL 5:	SET TIME
	18 ; ILD . . .100:	WAIT TIME
	19 ; CPN 5:	SET PART DOWN
	20 ; OCOF. . .1:	RELEASE PART
	21 ; TDL 5:	SET TIME
	22 ; ILD 100:	WAIT TIME
	23 ; CPN 4:	LIFT ARM AND PASS

INSTRUCTION	DETAIL
24 ; CPN 1:	GO HOME
25 ; NPN 9:	REPEAT

SUB ROUTINE 14	
BEGIN ; PGN 14:	ESTABLISH PROGRAM NUMBER
1 ; CPN 2:	GO OVER SPRAY
IMMED: 2 ; OCON. . . 3:	ACTUATE SPRAY
3 ; TDL . . . 20:	SET TIME
4 ; ILD . . .100:	WAIT TIME
5 ; OCOF. . . 3:	STOP SPRAY
6 ; SKIP. . . . 2: YES	CHECK FOR PART
7 ; SKIP. .1002: IMMED	GO BACK TO ACTUATE SPRAY
YES 8 ; NPN 14:	RETURN

List of Instruction Codes

Each instruction code is made up of two (2) parts; the function command and the function value.

Function Command. The function command is a general description of a task to be done, see Table 5-2.

Function Value. The function value is a specific description of what action should take place. An instruction code is used by the Control Unit to produce a definite action. The following is a list of instruction codes and their meaning.

Review Controls

The review controls are used to inspect each program sequence instruction contained in memory. The SINGLE STEP FWD and SINGLE STEP REVERSE Controls provide a means of executing one instruction at a time or beginning the program from a sequence other than the first instruction. In this way, the operator can review and inspect the operation of the program.

Single Step Forward. Once a program number has been selected and the mode select switch is placed in the SINGLE STEP position, pressing the SINGLE STEP FWD Control executes the first instruction in the program sequence. At this time, the program teach panel indicators will display the Program Number, Program Sequence, Function Command, Function Value, and the Position Number. When a new position is called, the JOG switch lamp will light. Pressing the JOG switch button will allow the robot arm to move

Table 5-2. Function Command Value Meaning.

FUNCTION COMMAND	VALUE	MEANING
CPN	XXX	GO TO POSITION (XXXX)
CSR	. . XX	GO DO PROGRAM (XX) & RETURN
CSR	NXXX	SET COUNT LIMIT (XXX) FOR COUNTER (N)
ILD	. . XX	WAIT FOR LINE (XX) TRUE
ILD	. 10X	WAIT FOR TIMER (X) TIME OUT
NPN	. . XX	DO PROGRAM NUMBER (XX) NEXT
OCOF	. . XX	DISENGAGE LINE (XX)
OCOF	. 10 X	DISABLE TIMER (X)
OCOF	. 2 XX	CLEAR FLAG (XX)
OCOF	. 30 X	DECREMENT COUNTER (X)
OCOF	. 40 X	DISABLE COUNTER (X)
OCOF	10 XX	IMMEDIATELY DISENGAGE LINE (XX)
OCOF	110 X	IMMEDIATELY DISABLE TIMER (X)
OCOF	12 XX	IMMEDIATELY CLEAR FLAG (XX)
OCOF	130 X	IMMEDIATELY DECREMENT COUNTER (X)
OCOF	140 X	IMMEDIATELY DISABLE COUNTER (X)
OCON	. . XX	ENGAGE LINE (XX)
OCON	. 10 X	ENABLE TIMER (X)
OCON	. 2 XX	SET FLAG (XX)
OCON	. 30 X	INCREMENT COUNTER (X)
OCON	. 40 X	ENABLE COUNTER (X)
OCON	. 50 X	DISPLAY COUNT FOR COUNTER (X)
OCON	. 60 X	DISPLAY COUNT LIMIT FOR COUNTER (X)
OCON	10 XX	IMMEDIATELY ENGAGE LINE (XX)
OCON	110 X	IMMEDIATELY ENABLE TIMER (X)
OCON	12 XX	IMMEDIATELY SET FLAG (XX)
OCON	130 X	IMMEDIATELY INCREMENT COUNTER (X)
OCON	140 X	IMMEDIATELY ENABLE COUNTER (X)
OCON	150 X	IMMEDIATELY DISPLAY COUNT FOR COUNTER (X)
OCON	160 X	IMMEDIATELY DISPLAY COUNT LIMIT FOR COUNTER (X)
ROP	XXXX	PASS THROUGH POSITION (XXXX)
SAD	. . XX	SET ACCELERATION TO VALUE (XX)
SKIP	. . XX	SKIP NEXT INST. IF LINE (XX) IS TRUE
SKIP	. 10 X	SKIP NEXT INST. IF TIMER (X) TIMED OUT
SKIP	. 2 XX	SKIP NEXT INST. IF FLAG (XX) TRUE
SKIP	. 30 X	SKIP NEXT INST. IF COUNTER (X) = LIMIT
SKIP	. 40 X	SKIP NEXT INST. IF COUNTER (X) ZERO
SKIP	1000 + XXXX	JUMP IMMEDIATELY TO SEQUENCE (XXXX)
SVL	..XX	SET VELOCITY TO VALUE (XX)
TDL	NXXX	SET TIMER (N) FOR TIME (XXX)

towards the position being displayed by the position number.
NOTE: It is recommended that the JOG switch button is operated for each position when a new program is first entered to the Control Units memory. In this way, the operator is assured that the robot arm has a free path to travel. If the operator is sure the robot arm can reach the position being

called for without striking any objects the RUN PROGRAM switch button can be pressed to move the robot arm to the new position at program speeds.

CAUTION: Once the RUN PROGRAM switch is pressed in the single stop mode, the robot arm will move directly to the position. Releasing the RUN PROGRAM switch button will not stop the motion of the arm. If under these conditions the arm is about to strike an object, the EMERGENCY STOP Control should be engaged.

Single Step Reverse. The Single Step Reverse Control is used to start a program from an instruction sequence other than sequence one. This may be desirable in the event an Emergency Shut down was required or power failure occurred, while the program was partially through its cycle. Starting the program from the sequence where the "stop" occurred to complete the operation that was in process, can be conducted as outlined in the following procedure.

Procedure:

(i) Place the Mode Select key in the SINGLE STEP position.

(ii) Operate the SINGLE STEP FWD Control until the desired program sequence number appears in the Program Sequence display.

(iii) Press the SINGLE STEP REVERSE Control.

(iv) Press the CONTROLLED STOP (yellow mushroom switch) on the Operator Panel.

(v) Press and hold the JOG switch until the robot arm is in the appropriate position.

(iv) Press the RUN PROGRAM Switch on the Operator Control Panel to execute the program beginning with the instruction sequence displayed in the Program Sequence Indicator.

NOTE: It is recommended that the COMPLETE PROGRAM THEN STOP Control is activated once the program begins operating in this mode. Once the program stops, the Mode Select key should then be placed in the AUTO RUN position the JOG switch pressed and held until the robot arm is in position 1; and the RUN PROGRAM switch activated to execute the program in the normal mode of operation.

Editing Controls

These controls are used to alter or change the program or instruction sequence when corrections are necessary. They consist of the INSERT and DELETE Controls.

Insert. The Insert Control is used to place an instruction in the program sequence. When the instruction is inserted at a specified program sequence number, the instruction originally at that program sequence number becomes the next instruction following the inserted instruction.

Procedure:

(i) Mode Select key to the TEACH position.

(ii) Press PROG SEQ NUMBER and the desired program sequence number on the keyboard: ENTER number.

(iii) Prepare the desired instruction code (function command and function value) to be inserted.

(iv) Press INSERT.

Delete. The Delete Control is used to take out the instruction specified by the program sequence number selection. When an instruction is deleted, all following instructions move up one number in the program reference to avoid gaps in the program. In this way, the next instruction, following the instruction deleted, will be executed in the program sequence where the deleted instruction had been originally.

Procedure:

(i) Mode Select key to the TEACH position.

(ii) Press PROG SEQ NUMBER and the desired program sequence number on the keyboard: ENTER the number.

(iii) Press DELETE.

With the use of Editing Control, the program can be modified without having to change the sequence number of all instructions.

Simple Programs/Problems

The following provides three (3) examples of programs which use many of the instruction codes contained in the Instruction Code List (Section E). Each program provides a description of the task and an explanation of the program.

Following the examples are three sample problems which can be used as an exercise for the reader. A solution to these problems can be found at the end of the section. However, other solutions are possible and the reader is urged to use imagination and ingenuity in solving the problem.

EXAMPLE 10.1:

(a) Task: To start at Home and move the robot arm to location A. At location A, set the horizontal arm into a stroking motion (back and forth) and repeat the stroke five times. Then move the robot arm to location B passing through Home and repeat the stroking motions of the horizontal arm at location B, return the robot arm to Home and repeat the process.

(b) This program calls for the same routine to be done at both locations A and B. The stroking motion of the horizontal arm can be put into a subroutine motion of the main program. Since subroutine motion of the robot arm is relative to the position of the arm at the point of departure from the main program, the positions A and B will become position 1 of the subroutine. It is further necessary to keep track of the number of strokes in the subroutine, since the task limits the number to five.

(c) Source Program and Detail:

Main Program

BEGIN ;		PGN . . 10:	REQUEST PROGRAM NUMBER TO BE USED.
	1;	SVL. . . 8:	SET VELOCITY TO MEDIUM SPEED.
	2;	SAD. . .8:	SET ACCELERATION TO MEDIUM RATE
	3;	CPN. . .2:	MOVE THE ARM TO LOCATION A.
	4;	CSR. .11:	GO DO PROGRAM NUMBER 11.
	5;	CPN. . .1:	GO HOME.
	6;	CPN. . .3:	MOVE TO LOCATION B.
	7;	CSR. .11:	GO DO PROGRAM NUMBER 11.
	8;	CPN. . . 1:	RETURN TO HOME.
	9;	NPN. .10:	REPEAT THE PROCESS

Subroutine

BEGIN ;		PGN. .11:	REQUEST PROGRAM NUMBER TO BE USED.
	1;	CSR 1005:	SET COUNTER TO LIMIT 5.
BACK:	2;	CPN. . .2:	MOVE ARM BACK.
	3;	CPN. . .1:	MOVE ARM FORWARD (RETURN HOME).
	4;	OCON . 301:	INCREMENT COUNTER
	5;	SKIP . 301: YES	IS COUNTER AT LIMIT 5?
	6;	SKIP 1002: BACK	JUMP BACK TO PROG SEQ 2.
YES:	7;	NPN. .11:	RETURN TO MAIN PROGRAM.

187

Program sequence 1 clears the counter (SET TO ZERO) and establishes a limit of 5 counts. Sequence 2 and 3 provide the single stroke motion. Sequence 4 adds one to the counter. Sequence 5 asks if the counter has reached a count of 5. If it has not, the next instruction (sequence 6) returns to sequence number 2 where the stroke and count are again executed. If five strokes have been counted program sequence 5 requires that sequence 6 is skipped, consequently sequence 7 instructs a return to the main program to the next instruction following CSR XXXX: instruction.

(departure point)

EXAMPLE 10.2:

(a) <u>Task:</u> Start at Home and go pick up parts at location A and move them to location B for a 5 second test. If the part tests good, move to location C and dump the part, if the part tests bad, move to location D and dump the part. Furthermore, if after 10 parts, two or more are bad, stop the process and sound an alarm. If less than two are bad, test 10 more parts.

NOTE: This program will assume the part pick up device is on line 1. The external TEST source is on line 2. A good parts sends a TRUE signal on line 2, and the alarm is on line 3.

(b) Source Program: See page 189.

EXAMPLE 10.3:

(a) <u>Task:</u> To get part at location A and move part to location B. At location b, give part a two minute bath with rotary motion; after two minutes, move to location C and dump part. Repeat the process.

NOTE: This process calls for an extended time interval beyond the 99.9 second limit of a timer. Consequently, setting a timer for one minute and executing the instruction twice will accomplish the desired task.

The program will assume the part pick up device is on line 1 and the part rotate device is on line 2.

(b) Source Program: See page 190.

Thus we conclude this study of how an industrial robot, the Prab Versatran may be programmed and controlled and taught. Other types of industrial robots will have a similar concept of operation. It is not difficult to program such a robot

Main Program

	BEGIN ;	PGN. .20:	REQUEST PROGRAM NUMBER TO BE USED.
	1 ;	SVL. . .8:	SET VELOCITY AT MEDIUM SPEED.
	2 ;	SAD. . .8:	SET ACCELERATION AT MEDIUM RATE.
	3 ;	CSR 1010:	SET COUNTER 1 to LIMIT 10 (Parts Tested).
	4 ;	CSR 2002:	SET COUNTER 2 to LIMIT 2 (Bad Parts).
BACK:	5 ;	CPN. . .2:	GO TO LOCATION A TO GET PART.
	6 ;	OCON. .1:	PICK UP PART.
	7 ;	OCON . 301	INCREMENT COUNTER 1.
	8 ;	CPN. . .3:	MOVE TO LOCATION B TO TEST PART.
	9 ;	TDL .50:	SET TIME 5 SECONDS.
	10 ;	ILD . 100:	WAIT FOR TEST TO BE COMPLETED.
	11 ;	SKIP . . .2: YES	PART BAD? YES; SKIP NEXT INST.
	12 ;	CST. .22:	GO DUMP GOOD PART.
YES:	13 ;	SKIP. . .2: YES 1	PART BAD? YES; SKIP NEXT INST.
	14 ;	SKIP 1016: JUMP	JUMP TO INST. 16.
YES 1:	15 ;	CSR .23:	GO DUMP BAD PART.
JUMP	16 ;	CPN. . .1:	RETURN HOME.
	17 ;	SKIP . 301: YES 2	10 PARTS TESTED? YES? SKIP NEXT INST.
	18 ;	SKIP 1005: BACK	GO BACK TO 5 and DO AGAIN.
YES 2:	19 ;	SKIP . 302: YES 3	2 OR MORE BAD PARTS? YES; SKIP NEXT.
	20 ;	NPN. .20:	REPEAT FOR 10 MORE PARTS.
YES 3:	21 ;	OCON. .3:	SOUND ALARM.

Subroutine (Good Part DUMP)

BEGIN ;	PGN. .22:	REQUEST PROGRAM NUMBER TO BE USED.
1 ;	CPN. . .2:	MOVE TO LOCATION C (GOOD DUMP).
2 ;	OCOF. . 1:	RELEASE PART.
3 ;	CPN. . . 1:	RETURN HOME.
4 ;	NPN. .22:	RETURN TO MAIN PROGRAM.

Subroutine (Bad Part Dump)

BEGIN ;	PGN. .23:	REQUEST PROGRAM NUMBER TO BE USED.
1 ;	CPN. . .2:	GO TO BAD PART DUMP.
2 ;	OCOF. .1:	RELEAST PART.
3 ;	CPN. . . 1:	RETURN HOME.
4 ;	. OCON . 302:	INCREMENT COUNTER 2 (BAD PART).
5 ;	NPN. .22:	RETURN TO MAIN PROGRAM

189

(b) Source Program: DETAIL

```
        BEGIN ;        PGN. .40:   REQUEST PROGRAM NUMBER 15
            1 ;        SVL. . .8:  SET VELOCITY MEDIUM SPEED
            2 ;        SAD. . .8:  SET ACCELERATION MEDIUM RATE
            3 ;        CSR 1002:   SET COUNTER 1 TO LIMIT 2
            4 ;        CPN. . .2:  GO TO LOCATION A (PART PICK UP)
            5 ;        OCON. .1:   PICK UP PART
            6 ;        CPN. . .3:  GO TO LOCATION B(BATH)
            7 ;        OCON. .2:   ROTATE PART
BACK:       8 ;        TDL . 600:  SET TIME 1 MINUTE
            9 ;        ILD . 100:  WAIT TIME
           10 ;        OCON . 301: INCREMENT COUNTER 1
           11 ;        SKIP . 301: YES  IS THIS TWICE? YES; SKIP NEXT
           12 ;        SKIP 1008: BACK JUMP BACK TO SEQUENCE 8
YES:       13 ;        OCOF. .2:   STOP ROTATING PART
           14 ;        CPN. . .4:  GO TO LOCATION C (PART DUMP)
           15 ;        OCOF. .1:   RELEASE PART
           16 ;        CPN. . .1:  GO HOME
           17 ;        NPN. .40:   REPEAT THE PROCESS
```

and the tasks may be varied a great deal from the illustrations used herein. Also, it is possible to use one computer to control many robots on an assembly line, or to integrate the actions of many robots to accomplish whatever jobs are desired.

Robot "Hands" and Arm Kinematics

We assume that a robot has at least one arm which can move, and that the robot can turn to any desired azimuth to perform the task which we have programmed into it. In the types of industrial robots we have considered so far, one arm per machine has seemed to be standard. Of course we are considering that a second arm, not attached to the robot itself, but controlled by the same master controller and synchronized with the movement of the main arm is a distinct possibility. However, in this chapter, we will deal with the "hand" or manipulator of just a single arm.

COMPARISON TO THE HUMAN ARM AND HAND

Because humans tend to think in terms of what they can do with their own appendages, they tend to design robots so that they duplicate humans. Somehow, in the back of our minds, we are thinking, "If we can do it, then a robot built like us can do it." So, we want the robot to be such a machine that it actually tends to duplicate our own movements with regard to the upper torso. As far as mobility is concerned, that doesn't seem to be much of a consideration except for the fact that the robot must be able to traverse the type and kind of terrain in which it will operate if movement is required. It is not necessary that a robot have legs and ankles and feet, nor

191

that it have just two of these if it is actually necessary that it duplicate the humanoid from the waist down, mobility-wise. In some cases where the terrain is rugged or mountainous more appendages have been considered and a spider-like bottom half is what the designers came up with. This, of course, would be especially necessary if the terrain is such that one cannot go around obstacles, but has to climb over them.

The robot must duplicate the human upper torso if the tasks assigned are such that it is doing work which a human formerly performed, and thus the rest of the work environment is geared to humans. Programming becomes easier because one can plan the movements and operations by comparing the robot's movements to those of a human who is very good at that job. Of course, one might take the viewpoint that a robot should be designed so that it is a unique device and *doesn't* duplicate the human.

We want to consider some of the similarities and differences between a robot and the human appendage. First let us consider the shoulder. In the simplest case the robot's shoulder joint would permit an up and down movement of the rest of the arm. To get the azimuth movement the human enjoys, the robot's body is turned in a circle. The result is very close to a human arm when the human is lying on their side. The human arm can describe a circle using the shoulder joint alone as the pivotal center.

The robot is capable, without stress on design, of an arm movement from the shoulder joint, which a human is not capable of; that is the extension or retraction of the upper arm elements. This permits the robot's arm to reach further and work closer than a human might be able to. Where a human might have to take a few steps to reach something, then step back to get close to the work area, the robot, by means of the extension capability, can remain in a fixed position and do the same task. The elimination of the mobility requirement here might be a decided plus for the robot in that it cannot get out of line, has fewer moving parts to worry about, and is simplified in design and cost. It is more cost-effective, in this manner.

Now we consider the elbow joint. It could be important that this joint have a dual capability, or triple capability. It should permit the forearm to lengthen and contract, it should permit up and down motion, and it should permit a rotary motion so that the "up and down" movement can be at any angle.

The robot, to have maximum manipulator movement, needs maximum flexibility of arm movement. If a robot is designed in such a way that it is to accomplish tasks which are not envisioned at the time of design, then a maximum capability of design has a better chance or permitting the robot to do these unforeseen or unimagined tasks later in time. Of course one realizes that for a specific task the robot under consideration may not have to have as much arm movement or flexibility of appendage as we are now considering. Less flexibility may mean less down time and certainly less cost. Thus one encounters, again, the concept of cost effectiveness of a robot doing one or two or a few tasks versus the completely flexible robot.

The wrist of the robot certainly needs a rotary capability, and an up and down capability. It probably does not need the extension capability of the other two arm elements. The wrist of the robot is where the gripper will be attached, and that device is what will enable the robot to do useful work. To do useful work the robot must have the right tool in the right place at the right time and manipulate that tool in the right manner for the required time to accomplish the task.

The human wrist has a good movement up and down and a somewhat limited movement sideways, but it can also be rotated, when considering the play of bones and muscles from the elbow joint down to the hand. If you examine your own hand's movement you'll find that the wrist rotates about 180 degrees only, but the hand can be rotated this much when bent down or up, or even, to some extent, when it is bent sideways. It is not the elbow which permits the rotary movement of the wrist, but the forearm bones and muscles, and the elbow doesn't cause any arm rotation at all, but we might want this in a robot. It is easier to design the joint which has both translatory movement and angular movement, then to try to exactly duplicate the bone structure of the human forearm. Remember, what we will be looking for in the final design, is movement of a hand or gripper through those necessary points in space required for accomplishment of whatever task we have assigned to that robot.

Finally, we come to the hand or gripper. Now we need to make some slight distinction. If we call the end appendage a hand then we want to consider that it has some kind of fingers or graspers in an amount of three or more, one of which might be considered a thumb or opposing lever. The gripper on the other hand might just be considered to be a pincers arrangement of two fingers, which close on an object to a predetermined force level, and permit the arm to raise or move the object so grasped.

The appendage attached to the wrist of an industrial robot may be some tool such as a drive socket device to fasten down bolts, or a drill, or a welding torch, or whatever. A good robot may be able to use one tool on a task, then move its arm so that that tool can be discarded or disengaged. The arm then moves to the location of another tool, moves the wrist so that the second tool is fastened in place on the wrist, and then the arm moves to perform tasks with the new tool. In this case there are no fingers or grippers as such. There is just the end tool which is positioned by the robot so that it can do the kind of work it was intended for.

A side note here to those of you who thought that fingers on a robot would be comparable to fingers on a human. Although possible, such a thing is probably not very useful. Take a look at Fig. 6-1. As you see, in this sketch, there are three finger sections shown which represent the end joint, the second joint, and the first joint of a finger. If we provide some small nubs at each joint we can exert a lever force on the finger element causing that part of the finger to move up or down as we would push or pull some flexible push-pull rods as shown. Such rods are ideally worked by small hydraulic pistons like the model airplane type used for landing gear extension and retraction. As some force is necessary, and the lever

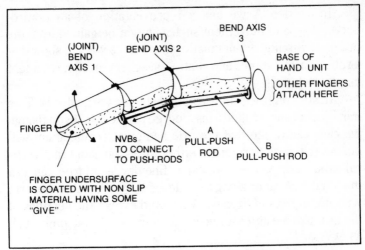

Fig. 6-1. Making a metal finger close and open using push rods.

advantage is relatively small, the hand would not have a strong grip.

If we imagine such a finger grasping something, we would imagine that push-rod B is pulled back moving the second finger element down, then push-rod A is pulled back causing the finger end element to come down and so we have a kind of grasping action. The push rods would have to be flexible enough to withstand the bending, or a two way cable arrangement might have to be used on each nub to get up and down movement of the finger element. Mechanical engineers love a problem like this and delight in imagining just how to design the mechanism to do the job as shown. But even if such an appendage is useful, it is doubtful if it is cost effective, so we leave it to your imagination.

SOME SCHEMATIC SYMBOLISM

When studying mechanical robots which have many moving parts it is sometimes useful to diagram those elements which move in a simplified manner. If you are so inclined, and talented enough, you can write some equations of motion which describe the moving elements. We examine this type of robotic symbolism in Fig. 6-2.

In section A we see a representation of an inward-outward type of movement such as might be gained from the use of a hydraulic or pneumatic piston. If we want to show that a robot's arm extends and contracts, schematically, we might use this symbol.

In section B we show rotation about a vertical axis. This movement can be called a one degree of freedom, even though we can readily see that the axle and pivot arrangement will permit the mass to rotate in either direction about that vertical axis. Perhaps then, we can define degree of freedom as a movement about or along a single axis. If we have a robot arm with *five degrees of freedom,* we would expect the sections of the arm to have at least five hinge joints, or thrust joints such as at A or B.

In C of Fig. 6-2 we see an arrangement which describes two degrees of freedom. Notice that the mass (M) can move around a horizontal axis or a vertical axis or both. In D we show a schematic of a free gyroscope which has three degrees of freedom. If we want to use our schematic representation to show this configuration, we do so as at E.

When the kinematic or dynamic situation of a robot's arm is being considered, it is nice to use such diagrams, adding one to another so that the entire picture may be obtained. The mass as shown does not represent the load which the arm may be moving, it can represent the inertia of the arm itself. That is a considerable problem in dynamics because the inertia of the arm changes with each position it assumes, with respect to most pivotal joints. This makes the equations of motion difficult to write and to solve!

Most robot arms currently have under six degrees of freedom. But, as an exercise let us examine an arm with 12 degrees of freedom, in Fig. 6-3. As a sure cure for insomnia one might imagine writing the equations of this arm, recalling that each degree of freedom involves a second order differential equation. Then imagine a method of simultaneously solving all the equations! If sleep doesn't come soon, consider the changes in inertia previously mentioned and some types of other non-linear relationships which have to exist in such an arrangement.

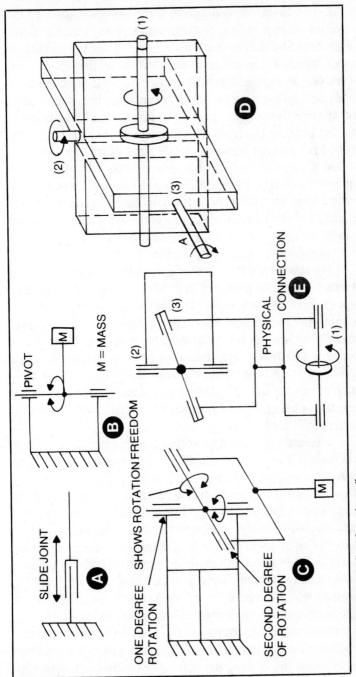

Fig. 6-2. Robotic symbolism for schematics.

197

It has been pointed out that the determination of the required number of degrees of freedom of any robot arm will be governed by the application of that robot and the conditions under which it must operate. Recall how the Prab robot overcame an obstacle in its arm trajectory? If we consider a robot doing a painting job inside a car body, in order to paint the interior thoroughly, getting the arm through a body opening and bending the hand around to spray the interior, might indeed require more than a few degrees of freedom. A General Motors painting robot has sensory capability to detect a car body entering its world, spray it carefully, and using a second hand open the door so the painting arm can get through to spray the interior! Coordination of the arms and hands is, of course, computer controlled. This robot uses both magnetic and optical sensors in its operation.

From what we have considered with respect to the arm of a robot, we now have a clear concept of the two things that happen. First the arm moves the end device into the required position in three dimensional space. Then the end device is actuated by signals from one or more sensors. When it has completed its task—determined by other sensors—it moves back to the starting position. Meantime the arm may have changed its spatial orientation as a function of time or its program sensing progress detectors, so that an entire job may be accomplished.

We need to mention two other types of end devices. The first is the sucker, which can hold onto various parts simply by creating a suction or vacuum. Normally those parts held would be rather light weight parts, although lots of suckers can produce a formidable attractive and holding force. We also need to mention the magnetic holding power to electromagnetic elements in hands or grippers. These exert considerable force which can easily be turned off and on by means of electric switching. They are not useful when handling non-magnetic type parts or when the parts are apt to have a high degree of residual magnetism, however.

Limit switches, which can be made an integral part of a hand are also very useful in that they can provide a feedback signal when the closing pressure has reached a given mag-

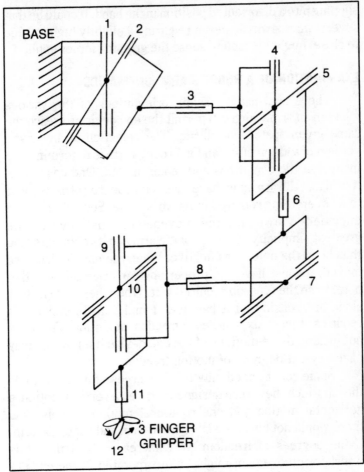

Fig. 6-3. A robot arm with 12 degrees of freedom.

nitude. These type sensors would be mounted inside the hand in such a way that when the hand clamps an object, pressure would cause the switch to close and thus create a signal that a given force had been exerted by the jaws. The system might be so programmed that once such a switch has closed, the controller would automatically cause a continued closing of some very, very small amount. A sensor tells the computer there is contact with the object, and the computer, knowing the flexibility of the object, orders a slightly tighter closing so that the object might be lifted without falling from the hand or slipping into a disoriented position in the hand. It could be that several microswitches each requiring a slightly higher force to close, might be used to sense the gripping forces applied.

EXAMINATION OF A ROBOT'S ARM KINEMATICS

There are many possible orientations of the various joints in an arm which can permit the end device to reach the same given spatial coordinate. The selection of any given version of a kinematic plan for a robot's arm is determined by many specific conditions and requirements. One consideration is that the arm must have a sufficient degree of movement to do everything required in its work cycle. Second, from the engineers viewpoint, the movements must involve the greatest simplicity of design of the arm and its elements, and thus keep the unit cost effective. When the end device is so moved, it must have the proper angular orientation at that point to do the type job required of it. When these two factors alone are considered it has been found that, in general, it requires at least six degrees of freedom of motion. This does not include those motions of the end device itself which may have several degrees of motion freedom.

Some robots need only three degrees of freedom to do the job which they were designed for; i.e., a vertical motion, a horizontal motion and one rotational motion. In this case there would not be a translatory motion in and out of the arm. If the degrees of freedom were increased to 4 then this translatory movement might be incorporated. If the end device must have a certain number of degrees of freedom, the arm must make up the number of degrees of freedom which the end device does not produce or have in itself.

Since, we know that the number of degrees of freedom can represent various kinematic plans for the arm. The designer can lay-out the kinematics of the arm to satisfy such requirements as he may deem necessary for that particular application of that particular robot. Of course, he may over design the arm for some installations and realize that the arm will just meet the requirements for another installation, but this gives the arm more use in a wider spread of applications, even though it becomes less cost effective for the smaller application.

When a designer considers the arm itself and its kinematics he will want to consider several factors. One of these is that the change in the angular orientation of the end device when it is at some specified point in its work zone, must not involve significant movements of the arm elements which are located some distance away from the end device. One brings this down to an earthy level by saying that when a human wants to grasp a pencil, he moves his fingers, not his shoulder muscles. The designer considers the distribution of the necessary number of degrees of freedom when the total number of freedom degrees is a constant. If the cost effectiveness requires that you can have only four degrees of freedom, where should they be located?

So we find that in order to maintain the necessary maneuverability of a robot's arm seven to nine or more degrees of freedom might be required on some tasks. You don't have to play around with mathematic or proofs to show this, just layout the work area and draw the kinematics required for the tasks.

POWERING THE ROBOT'S ARMS

General Motors expects to use about 14,000 industrial robots in their factories by 1990. As many of the parts and tools used in automobile production are quite heavy, we need to further examine a robot system with regard to how the arm is powered.

At each joint, or section of the machine, there must be powering units to control the movements and position of the extensions appended thereto. We might even conclude that a robotic arm consists of many individually controlled elements, each with its own set of operating signals and sensors, which become integrated into one unit under the control of a master computer, or programmed drum, or both. The drive devices, which we are at the moment interested in, are placed in various ways to control the motion of the arm. Most generally they are placed directly at the joints or on the elements of the arm near a joint. But because you don't get much leverage, this is suitable only when the arm is not required to lift heavy loads. We must also consider that type of drive where the powering units are all located near the shoulder joint and use other means of transmission of power from the drive sections to the elements of the arm. In some applications this may be a good way to power the elements so that the weight of the drive units are not on the arm itself, and thus the weight lifting capability vs the weight of the arm ratio might be increased. Finally, we must consider the third alternative, which is that there be a combination of element-drives and remote-drives for the operating machinery.

The three types of drives used are the electric, the hydraulic, and the pneumatic. Each has some particular advantages when used in robotics applications. The pneumatic and hydraulic systems require pumps at some remote location, with only the hoses going to the machine. Control can only be precise using sensors, however, load capability is excellent. Electric motors are another matter.

The electric motors are easily powered and may be designed for low inertia. Thus they are capable of excellent acceleration, deacceleration, and velocity control, but they are heavy for the amount of horsepower produced. They have excellent reliability records. In the past there have been some machine operations using electric motors working at constant speeds, using various types of speed reduction units, which may be controlled electronically, and also using clutches to transmit the power from the drive unit to the various elements. Clutches can engage and disengage quickly and have, in the past, offered quite high reliability in such applications. It is possible that for certain robotic applications in the future, some systems or even parts of systems, may use clutches and electric motors as a part of their powering system.

Without clutches it is necessary to use stepping motors in robots. Robot manipulators using these powering devices are capable of very discrete and tight control and such motors work easily with digital type input signals. If the stepping movement is large, then some auxiliary type equipment needs to be used, on the grippers to turn this step movement into a smooth precise movement of the end device.

Finally, we should not overlook the use of such items as a powered selsyn type of transmission and powering system for certain robotic applications. These may have either a physical or electrical input and can be very precisely controlled even though their physical power output is very low. They might be used in conjunction with other types of powering units and so give the smoothness and precision and power required for many applications.

THE PLANET ROBOT "ARMAX"

In Fig. 6-4 we see a one arm robot made by the Planet Corp. of Farmington Hills, Michigan. The grippers may be replaced with other tools as required. This unit uses hydraulic power and the source of that hydraulic power is usually remotely located from the unit. This robot is designed to be used in welding, material handling, machine loading/unload, gaging, deburring, flame cutting, surface finishing, and spray painting operations. Figure 6-5 shows its world. In this illustration we can see the vertical and horizontal profiles of its work area.

In Fig. 6-6 we see how the robot is used to manipulate a wire brush over an automobile body. The brush must come down on the body at a given point and with a given pressure. Then it must move precisely over a pre-planned curve, which might be a welded seam, for example, and then at the end of its movement, it needs to lift off the body and wait until another body slips into position. The machine must be taught just how to move its brush, of course, and a computer will memorize that information.

Planet Corporation says their robots may come in two variations; a servo controlled robot or a programmed robot. The servo-controlled type is operated by a computer which has one section for each controlled axis of movement. It uses a high level language and incorporates some adaptive control so it can adjust to changes in its external environment. This robot is taught with a hand held unit similar to the ones we have previously described.

Fig. 6-4. The Armax robot of the Planet Corp. (courtesy Planet Corp.).

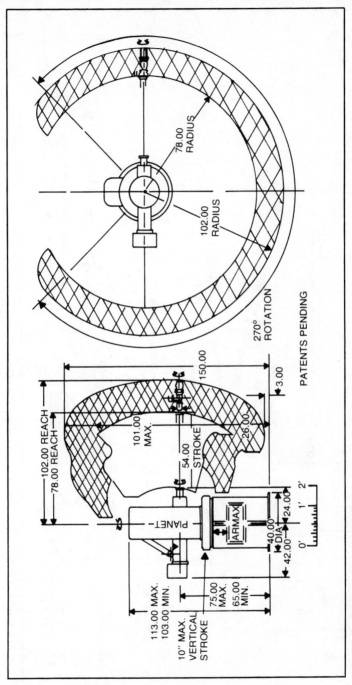

Fig. 6-5. The World of Armax, the robot.

205

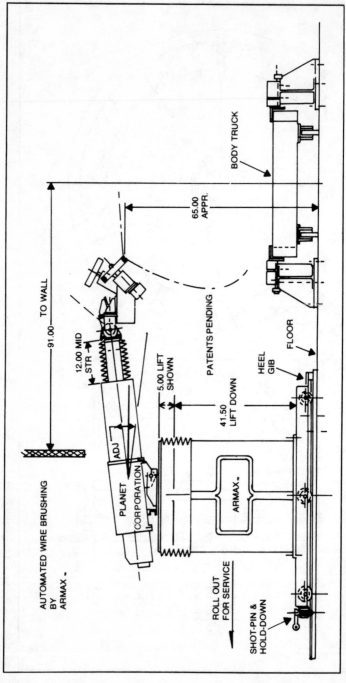

Fig. 6-6. Armax robot operates a wire brushing wheel (courtesy Planet Corp.).

In Fig. 6-7 the layout of the Armax robot is shown and its operation is indicated for a parts handling job. This job consists of handling a part which must be deburred and then placed in a different location. This is a programmed robot. That means that its actions are controlled by pegs in a drum which close various switches causing the various operations to be performed. We look at note 1. The spacer grippers open, the arm extends the grippers close and the arm retracts with the item and turns clockwise. In position 2 the arm extends, the machine opens up, the grippers open and release the part to the machine. The arm holds still while the de-burring procedure takes place, then the gripper closes and the arm retracts again and the robot arm is ready to rotate clockwise. In 3 the robot arm rotates into the position indicated and some tooling grips the part and the robot arm is ready to rotate clockwise. In 4 the robot arm rotates into the position shown, the arm extends and the tooling ungrips the parts. Now the arm retracts and the arm rotates counterclockwise to start at position 1 again. Notice that a gravity feed is used on the parts, and also that the parts are fed by a mechanical escapement which is operated as the arm rotates. This is necessary because control is not accurate enough to pick up and precisely orient small parts with a hydraulic robot.

Figure 6-8 illustrates the positions the arm takes when it is used in a machine load/unloading operation. Notice the legend which uses a letter P for pick-up, and the letter D for deposit of the part at the point indicated by the numbered arrow. Now, let us take a look at hydraulic drive and control.

THE HYDRAULIC DRIVE SYSTEM FOR ROBOT ARMS

The basic piston—transfer-valve drive for a hydraulic system is well known and the equations are easily derived when considering the flow rate, dimensions of the piston and valve, coefficients of friction, orfice flow, and the mass and load acceleration requirements. The basic unit and equations in a simplified form are shown in Fig. 6-9. In this diagram we see that the total force exerted by the piston will be equal to the differential pressure (p1—p2) and that this pressure will, in turn be governed by the flow rate of oil through the transfer

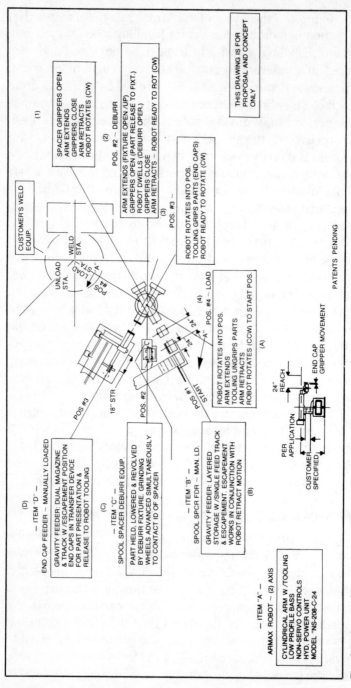

Fig. 6-7. Armax does parts handling. Its method of operation (courtesy Planet Corp.).

208

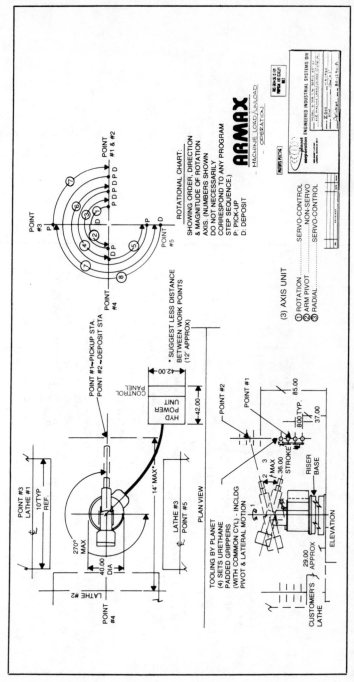

Fig. 6-8. Armax loading/unloading capability (courtesy Planet Corp.).

209

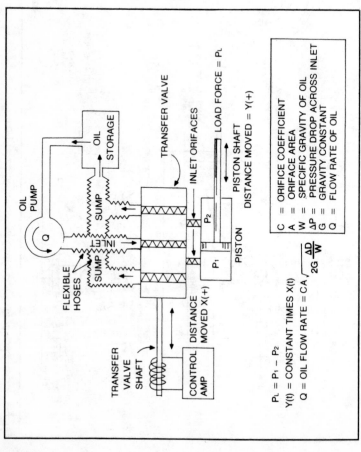

Fig. 6-9. The hydraulic piston-transfer valve diagram and equations of operation simplified.

$P_L = P_1 - P_2$

$Y(t) = $ CONSTANT TIMES $X(t)$

$Q = $ OIL FLOW RATE $= CA\sqrt{2G\dfrac{\Delta D}{W}}$

C = ORIFICE COEFFICIENT
A = ORIFICE AREA
W = SPECIFIC GRAVITY OF OIL
ΔP = PRESSURE DROP ACROSS INLET
G = GRAVITY CONSTANT
Q = FLOW RATE OF OIL

OIL PUMP

Q

OIL STORAGE

FLEXIBLE HOSES

SUMP

INLET

SUMP

SUMP

TRANSFER VALVE

INLET ORIFACES

LOAD FORCE = P_L

PISTON SHAFT
DISTANCE MOVED = $Y(+)$

P_2

P_1

PISTON

TRANSFER VALVE SHAFT

CONTROL AMP

DISTANCE MOVED $X(+)$

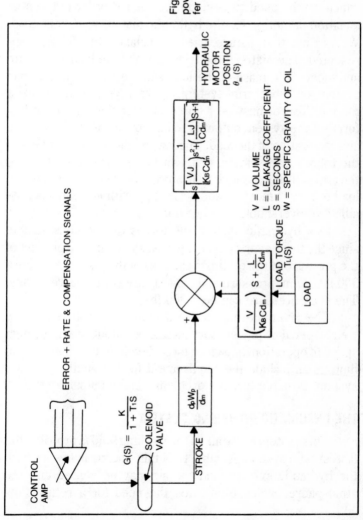

Fig. 6-10. Block diagram of hydraulic power actuating system, with transfer functions specified.

V = VOLUME
L = LEAKAGE COEFICIENT
S = SECONDS
W = SPECIFIC GRAVITY OF OIL

211

valve into the piston. We note that the rate-of-change of the piston shaft position (y) turns out to be some constant times the transfer valve shaft displacement (x).

A little more elaborate diagram which shows the control amplifier with its many inputs, the solenoid, and a load reaction. The expected transfer force minus the load reaction equals the actual transfer of power. The arm position sensor feeds information back to the control amp.

Finally, we show in Fig. 6-11 a very simplified diagram which can be used to develop the second order differential equation of motion for the hydraulic piston, valve, and load. We realize that nature tries to balance the forces represented. The piston force tries to make the load accelerate and move. The mass of the load and the friction in the load system or robot arm system or whatever, tend to resist motion. There must be a force greater than the retarding forces if the system output is to be moved, and there is to be an acceleration of the load. The load might be the members of the robot's arm system, or might be the arm plus whatever the arm is lifting or moving. In any event we want the arm and load to move smoothly and accurately without vibration, oscillation, overshoot, or other minor problems.

In a hydraulic system the forces can be considerable since the force exerted on the piston shaft will be a product of the pressure of the oil times the area of the piston. Assume a 300 lbs/sq.in. pressure and a piston area of 6 square inches. The shaft pressure can be 1,800 lbs.

The favorable aspects of a hydraulic system are low weight per-unit-power, low inertia of output member, high speed of operation, and large force development. Some of the limitations include the requirement for the auxiliary power system, possible leakage problems, and minimal control.

THE PNEUMATIC POWERING SYSTEM

This system using air as the working fluid is probably the fastest type system possible to obtain. Inertia is low, as with the hydraulic system, and the air may be vented into the atmosphere, which eliminates the need for a return flow

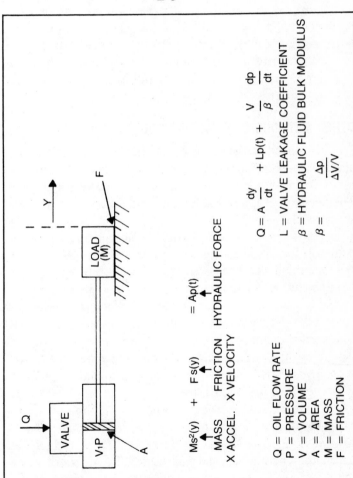

Fig. 6-11. Second order differential equation of hydraulic piston system.

$Ms^2(y) + Fs(y) = Ap(t)$

MASS   FRICTION   HYDRAULIC FORCE
X ACCEL.   X VELOCITY

$Q = A\dfrac{dy}{dt} + Lp(t) + \dfrac{V}{\beta}\dfrac{dp}{dt}$

L = VALVE LEAKAGE COEFFICIENT

β = HYDRAULIC FLUID BULK MODULUS

$\beta = \dfrac{\Delta p}{\Delta V/V}$

Q = OIL FLOW RATE
P = PRESSURE
V = VOLUME
A = AREA
M = MASS
F = FRICTION

213

system. Air is obtained from the atmophere and compressed into pressurized tanks.

In pneumatic systems one uses air under pressure to operate pistons, and also one may use the suction. This may be used to provide power to suckers which might be placed on grippers to hold certain types of packages or objects, or to attract them into the gripper jaws.

Although air is light weight and easily available, it poses some problems in that any contaminants in the system, even moisture, can cause trouble with system parts such as valves and tightly sealed pistons. But when its application fits a particular need, then it is used to quite a large advantage. On letter sorting machines used by the Post Office an arm moves to a stack of letters, grips one by its face with suction, moves it to another position by rotation of the arm, and then releases it with a short explosion of air.

OTHER METHODS OF POWERING ROBOTIC ARMS

The use of action envelopes with robotic drives has been studied by some. These are said to be quite sensitive and can range from low-powering to quite high powering-devices. They consist of tubes and bellows and follow the traditional action of these type elements.

You have a tube into which you can force non-compressible liquid such as hydraulic systems use. The tube does not lengthen, but tends to expand sideways and contract lengthwise, just as human muscles do. Thus a pulling action is available just as with a human muscle. When the pressure is released, the tube lengthens again and the sides return back to a normal diameter. No moving piston is necessary in this case. One attaches the load to the end of the tube and can pull it or, by use of a second tube and proper dynamics, push it to whatever position one desires.

Regardless of how the arm is powered, robots all use the same kinds of grippers and in Fig. 6-12 we show some developed by FIBRO-MANTA. They too are powered by electric motors, hydraulics, and pneumatics. The grippers may be caused to operate by proximity detection or by program. The use of springs gives a snap action which might be necessary

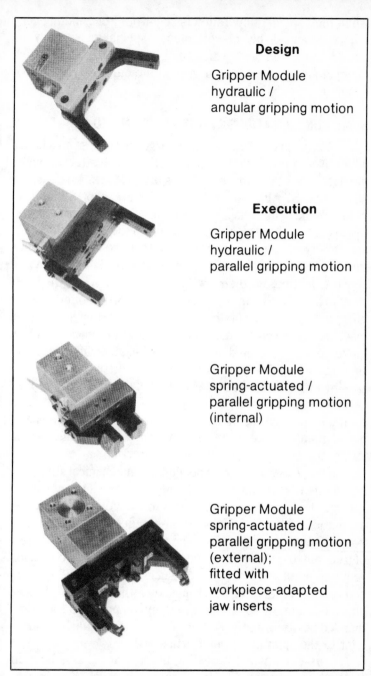

Design

Gripper Module
hydraulic /
angular gripping motion

Execution

Gripper Module
hydraulic /
parallel gripping motion

Gripper Module
spring-actuated /
parallel gripping motion
(internal)

Gripper Module
spring-actuated /
parallel gripping motion
(external);
fitted with
workpiece-adapted
jaw inserts

Fig. 6-12. FIBRO-MANTA gripper units for robotics (courtesy Fibro-Manta).

215

when high speed of operation is required. For handling certain types and shapes of work pieces, special inserts may be placed inside the jaws which then conform to the required dimensions to insure a tight grip on the piece of work being handled.

SOME CONSIDERATIONS OF THE PROGRAMMED ROBOT

There exist in our world a large number of machines which have been called robots. Many find this terminology not at all acceptable for these kinds of machines. As stated by Dr. James S. Albus of the National Bureau of Standards, "Most of these mechanisms exhibit few of the characteristics the average person would associate with the term robot; they are mostly pick-and-place machines that are capable of only the simplest kinds of motions. They have little or no ability to sense conditions in their own environment. When they are switched on, they simply execute a pre-programmed sequence of operations. The limits of motion for each joint of the machine are fixed by mechanical stops, and each detail of movement must be guided by means of an electric or pneumatic impulse originating at a plugboard control panel.

In a typical application, each row of the plugboard represents one degree of freedom in a specified robot joint, and each column represents one program step. Connections are made in appropriate rows and columns to determine which joints are actuated, and in which direction, at each step. Whenever a new program is needed, the mechanical stops can be re-positioned and the programming connections can be relocated to set up a new series of movements. A more sophisticated level of control can be achieved by adding servomechanisms that can command the position of each degree of freedom to assume any value. The addition of servo-control requires feedback from sensors such as potentiometers encoders and resolvers, which measure the position of each joint. The measured position is then compared to the commanded position, and any differences are corrected by signals sent to the appropriate joint actuators.

This excerpt is interesting because it tells us how, in general, a programmed system operates. A robot's arm may be moved by command signals which originate either from a timing and sequence drum which closes switches at various times, or from a computer which draws upon its memory for the movement commands. The difference is that the computer may have solved some equations as part of its own intelligent operation, it directly commands the movements and speed of the arm, and it analyzes the feedback to make adjustments. The program drum—which was set and patterned by us—could only move the arm one way, at one speed to get to its final position.

THE PATCHBOARD PROGRAMMED ROBOT

It really is a pretty dumb kind of machine. Let us examine Fig. 6-13. Please realize that this is just an example to illustrate our points. You can probably easily imagine a much better and more complete programming unit for a real-life job.

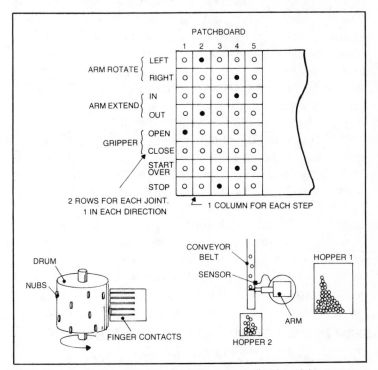

Fig. 6-13. A sketch of an imaginary "patchboard panel" which might program a robot.

Plugs go into the patchboard where the dark dots are. As a rotary switch makes contact with the plugs at each step, the corresponding joints are activated. The actions are stopped by mechanical stops or sensors attached directly to the arm.

This board will open the grippers, dropping an item into hopper 1, then pull in its arm as it rotates to the left. Positioned over the conveyor, it waits to close the gripper until the sensor is triggered. Notice this is *not* done by the patchboard. The sensor could sense light, or color, or, using magnetism, metal composition. The result is that items not tripping the sensor go into hopper 2. When the sensor is triggered, the grippers close and the rotary switch starts again, proceeding to step 4 where the arm moves out and right, over the 1st hopper. The START OVER plug indicates to the rotary switch (not the arm) to disengage and swing around to position 1, where it all starts again.

In some types of robotic machines a rotating drum is positioned at a starting point and around its circumference are such holes as we have shown and switch actuating pegs are placed in these holes. When the whole unit is placed into operation the drum starts rotating and closing and opening switches according to the various peg positions. The speed of the rotary switch or drum controls the period of time between steps, but the speed of movement cannot be controlled.

Although it may seem that this type machine is not useful, that is not true. It is simple and reliable and will perform its operations, even rather complex ones, with the same precision and accuracy time and time again. You know how useful and reliable the washing machine, dryer, and dishwasher are in your home? This kind of machine is that type. For some industrial operations this type machine may be much more useful and cost-effective than the servo-controlled machine. We will probably see everything from pencil sharpeners to lawn mowers called robots in the future years.

SERVOMECHANISM DEFINITIONS AND CONSIDERATIONS

The world of electro-mechanical control states that servo-systems fall into two general categories: the open loop system and the closed loop system. It is common to find the following definitions used with regard to these systems:

open loop system—A system of operation where the input does not depend on the output.

closed loop system—A system of operation where the input does depend on the output.

Examine Fig. 6-14. In Fig. 6-14A we see a typical open loop system which has no feedback. The output is expressed simply as the product of the various transfer functions of the various blocks of the system, as illustrated. These functions are usually expressed in Laplace Transforms.

Systems such as this open loop system can be used for control purposes assuming that we know where the arm is currently located, or that the arm is always returned to home or starting position at the end of each programmed operation or cycle. The input could be a series of polarized digital pulses sent from a computer that remembers where the arm is and where it needs to go next. The polarized pulse would turn a stepper motor the proper direction and distance. Assume that we have one such servo-system for each element of the arm. This means that if we have the shoulder section, the forearm and the wrist, that we would have three such servos in operation.

Our three servos might operate in this manner. The first servo will move the whole arm—it being the shoulder servo. Once the whole arm has been moved, then movement is relinquished to the forearm servo for more precise positioning of the wrist and gripper, or hand, and finally, the wrist is positioned so that the torch, paint gun, or whatever is in the desired position for task performance.

If we know the exact end-position that the hand has to reach, we can provide the correct number and type of digital pulses, as a function of time, to make the robot's arm move to where we want it to move. This concept of programming would work with the drum or patchboard too, if the servos have distances pre-set. Notice that no feedback is required. The whole machine simply gets an input of pre-arranged and polarized digital pulses which it amplifies and sends to vari-

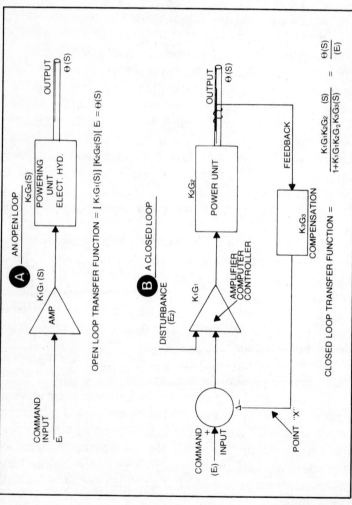

Fig. 6-14. Some concepts of Servomechanisms. The open loop and closed loop systems defined.

A AN OPEN LOOP

COMMAND INPUT
E

K_1G_1 (S)

AMP

$K_2G_2(S)$

POWERING UNIT ELECT. HYD.

OUTPUT
$\Theta(S)$

OPEN LOOP TRANSFER FUNCTION $= [K_1G_1(S)] [K_2G_2(S)] E = \Theta(S)$

B A CLOSED LOOP

COMMAND INPUT
(E)

DISTURBANCE
(E_2)

K_1G_1

AMPLIFIER COMPUTER CONTROLLER

K_2G_2

POWER UNIT

OUTPUT
$\Theta(S)$

FEEDBACK

K_3G_3

COMPENSATION

POINT "X"

CLOSED LOOP TRANSFER FUNCTION $= \dfrac{K_1G_1K_2G_2\ (S)}{1+K_1G_1K_2G_2K_3G_3(S)} = \dfrac{\Theta(S)}{(E)}$

ous stepping motors at a proper time so they will operate correctly and without mutual interference. This machine can also do useful work, accomplishing jobs suited into this sort of mindless type operation. The advantages of this type system are simplicity and low cost, because no feedback is needed, and a higher degree of accuracy than hydraulics are capable of.

In connection with the programmed movement of the arm of this type robot, there might be external mechanical switches operated by the arm as it passes. In this manner, bins might be opened as the arm swings toward them. Parts would then be fed into a belt. The combination of operating pulses of an electronic nature and the genius of mechanical engineers combine to make a machine do almost wondrous tasks! It seems there is a place for the open loop type control system in actual industrial applications.

Why, then, do we want a closed loop system? Using feedback and some internal computing element, it is capable of adaptive control. Only a closed system can adapt to varying situations.

It is interesting to note how the control equations of such a system are derived. One simply takes the product of all error detectors and then writes this group of terms as a fraction. Consider the loop opened at point X in Fig. 6-14. We can derive the open loop type transfer function to use in the formulation of the closed loop equation as shown. Realize, of course, that the KG terms can be very long and complicated expressions using the Laplace Transforms. Any good text on Servomechanisms will illustrate how this is done.

SERIES AND PARALLEL SERVOMECHANISMS

A system of series servomechanisms which might be used to control the arm of a robot is shown in Fig. 6-15. Assume that there is a command input from some place in the system. This causes the shoulder servo to activate, moving its output element physically until it is as close as it can get and the command is zeroed out. At the same time that it moves, it causes a command signal to the forearm element so that it also moves just as soon as the input command signal is large enough to cause the second servo system to operate. It,

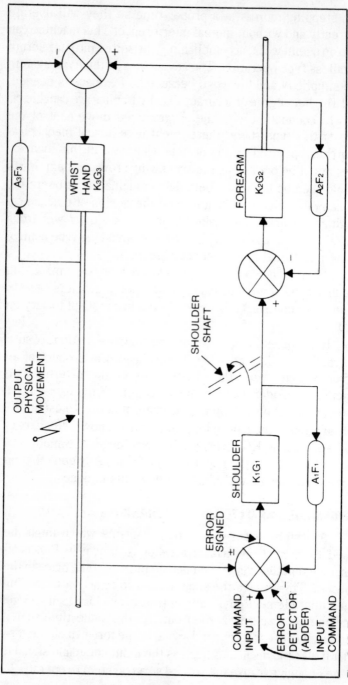

Fig. 6-15. The concept of series servos in a robotic system.

too, will move until the command signal is reduced below some operating threshold value—when the error is so small the servo sees no signal. Finally, the third servo is activated in a like manner and it moves until its input command is reduced below threshold value. In this movement, just to insure overall stability of the series arrangement, a feedback is made from the final output element to the primary input error detector, and this signal may be positive or negative as required for accuracy and stability.

Notice that we have a feedback around each servo for its own stability and precise operation, and that there will be an over-all loop from the output back to the input. Thus we have loops inside of loops! Techniques for writing the equations of such a system are well known. Computers can solve such equations and will permit playing around with variables, so one might have an interesting time seeing just what would happen in this situation, with and without load, with varying amounts of inertia, and if one considers some velocity and acceleration response situation requirements.

The input command, which may come from a computer, will govern the actual movement. The inter-relationship and inter-dependence of one arm element on its preceding element means that either the computer will have to do the calculations, or there will have to be pre-set servos, summing amps, or some other form of controlled feedback.

Parallel arrangement of servosystems can move the arm elements either together or sequentially as necessary. This situation might be diagrammed as shown in Fig. 6-16. Here we see how a servosystem controls the movement of the shoulder element. This servo has its normal feedback for stability and compensation, and it also provides feedback (1) to the computer driver which generates the input signal (A). Thus the shoulder element can move such that its end point (the elbow) will position itself at some coordinate as commanded by the input. The computer will know this and because of the feedback, will know just how the element is moving in space to get to that point. Meantime the computer can direct the second servo to move the forearm at the same time and along some pre-computed trajectory, and it can keep

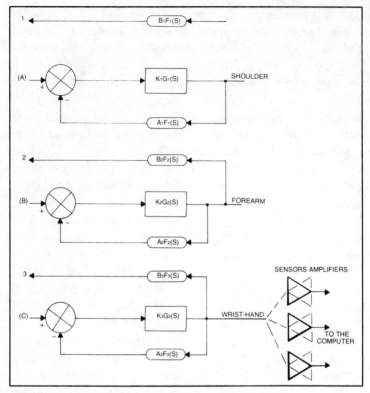

Fig. 6-16. Using parallel servos to control a robot's arm.

track of what is happening with this second element even as it programs and commands the first element. The same thing happens to the hand. Sensors on the hand will tell the computer that the hand is on the object, has grasped it, or that the hand is now ready for some further signals.

Just one further point concerning servomechanisms of this type. We have mentioned the threshold signal level. The smaller the signal which can make the servo operate, the smaller the actual error in position, velocity, or acceleration will be, however it is also more susceptible to noise.

ROBOT HANDS AND HUMAN HANDS

A study was made of the technical feasibility of constructing a robot hand which would be like a human hand. It was found that the human hand is just too complex for such duplication in the first place, and in the second place, it really isn't necessary to exactly and completely duplicate the human hand to have that robot perform well enough to replace humans at some tasks.

Some of the interesting facets of this study were that the human hand has some fifteen joints which result in some twenty two individual degrees of freedom. Think about writing equations to describe such movements! Although this number of degrees of freedom might be required for playing a piano, who wants a robot to play such an instrument? If you can't play it yourself, you can buy a player piano or record player.

The study showed that of the technically realizable manipulations possible for a hand and fingers, a hand having only one finger can perform about 5 percent of all the manipulations needed. If two movable finger parts are used (a gripper) then the totality of possible manipulations is increased to about 40 percent. If a third freely moving finger is added, with the same number of degrees of freedom as the human hand, then almost 90 percent of the possible technical manipulations can be performed. Adding a fourth finger increases the manipulative ability to some 99 percent. There isn't much use going beyond this capability.

If a two finger hand is our familiar gripper, the three fingered hand would be one in which fingers would be mounted on a wrist at 120-degree spacing, as in Fig. 6-1. If one were to design a hand with an additional rotary joint incorporated at the base of each finger, so that each finger could be rotated with respect to the others, and still maintain their gripping power and capability, then some 75 percent of all the manipulations of a human hand might be accomplished.

One "ideal" hand has four joints per finger, for three of the fingers, and the bases of these fingers are separated more than they are on the human hand. Three of the joints are controlled in whatever manner is most suitable, the fourth joint is automatically controlled by the second joint. The wide base arrangement permits the "thumb" to fold out of the working area between the other fingers. With this device some 89 percent of human hand operations might be achieved.

In a human hand the joints use sliding surfaces which, in part, are deformed during the motion of the joint. Also the human hand has very complicated gliding joints which are continuously lubricated with the joint fluid and have an exceptional safety system which prevents joint damage. The safety system consists of force and pressure sensors. The manipulating force per finger of the human hand is somewhere around 3.5 to 5 kg and is about 25 kg for all fingers working as a unit. It would be difficult to duplicate this variable capability by machine. With the pincers or gripper type hand one can design for large forces, up to 10 times the force of the human hand, and this type design has become very practical. The mechanical hand can operate at a faster velocity and acceleration than the human hand, and does not tire when moving at a fast pace. Also the wear on the mechanical parts is very small even when doing a fast task and exerting considerable forces in the task.

The mechanical hand may, or may not have environmental sensors. Infrared beams of light, generated within the grip of a mechanical hand could permit the hand to sense the proximity of objects it is trying to pick up. The human hand has many more sensors. It has a cold and hot sensor, feel sensors, pressure sensors, and force measurement sensors which are an integral part of the human hand's construction. If we include heat and cold sensors in a robot's hand, these might generate alarm signals if things begin to get too hot or too cold. Force measurement sensors are currently available in a number of different configurations, from limit switches and variable magnetic-signal transformers to piezo crystals. There are several hundred thousand receptors distributed over the human hand. Considering this, it is immediately obvious that we can never build an artificial hand that duplicates the human hand precisely! One hand has fingers which each have 12 degrees of freedom. The hand could give a rough idea of its surroundings by detecting corners, and determining structure symmetries, and so on without seeing the objects. The sensor signals were sufficient for a computer to build the image.

When humans solve a grasping problem, they are only

conscious of the final process which is to feel the object. They are no longer conscious of the individual motions which lead to the final result. This is all programmed by means of a lengthy learning process. Small children learn to grasp objects with their hands and do this at various distances from their bodies. They learn how to manipulate their hands and arms and wrists for round, thin, long, short, and fat objects. When this learning phase is over, they have a fixed program in their nervous systems or in their brains for the control of the various motion elements which make possible the manipulations required, and in such a manner that certain motion chains will automatically occur in a fixed sequence or desired sequence, and in synchronization with other parts of the human system.

The total robotic manipulation process is first decomposed into signals in a first, upper level. These signals determine which elements should be activated. The individual motion elements themselves are then activated in a second logical process. The final manipulation is produced by means of an additional synchronization control system governed by the adjustment of the pressure forces and the manipulation of these forces among the various body elements. The organization of such subordinate functions are autonomous, but influence each other in an adaptive control system manner. This is called hierarchically organized information processing. Much of the "feedback" information in robots is accomplished with analog circuits whose signals are then digitized for computer processing as we know.

CLOSED-LOOP SERVOSYSTEMS AND FEEDBACK

We have again mentioned feedback in connection with a closed loop type of servosystem. We shall see in later pages how a velocity feedback is important to the overall stability of one type of industrial robot. Let us examine the concept of feedback further in preparation for the information yet to come.

We can diagram a rather complex multiloop servosystem as shown in Fig. 6-17. There are three feedback loops shown, two of which have some kind of compensation amplifiers in their lines. The feedback to the input is, of course, the output shaft or element's position.

We define a transfer function as the ratio between output and input. The abbreviation for the transfer function is T.F. as shown in the illustration.

By being able to write a mathematical expression which tells us what happens to the voltages, currents, or physical elements controlled by the blocks, using the inputs and feedbacks as shown, we are able to examine how the output responds to changes in loading, gain of various blocks, the addition of various rates of change in voltage or current. Signals representing the velocity or acceleration of physical elements will affect the output, whether these signals develop in the feedback loops or in the forward part of the loop or block input line.

Because the computer will show the effects of these variations and also show whether to increase the torques or power applied, or increase the damping—which prevents oscillation—one can design a whole industrial robot system on paper and try it out on a computer before anyone has to spend a dollar buying expensive motors and gears and such, to make the thing, and try it out experimentally. Of course the final proof of design is always the building of a model and making such adjustments to it and corrections to its control as are necessary to make it do exactly what it is intended to do efficiently and precisely, and, hopefully, economically.

If we look at a general second order differential equation in the form $Jp^2x + Fpx + Kx = C$ or (Inertia × Acceleration) + (Friction × Velocity) + Power = C, where C is the command input. Then, if we know that the addition of a type of control such as velocity feedback will affect the second term or the damping term of the system, we can predict what will happen if we add this control to the system. Also, if we use a proper acceleration type feedback, we can predict whether this will have the effect of increasing or decreasing the system inertia. We can vary the value of the system power term (K) and see what happens. In a general way, we can add anything to the block diagram, write its transfer function, mathematically manipulate the equation till we get it in the form shown and

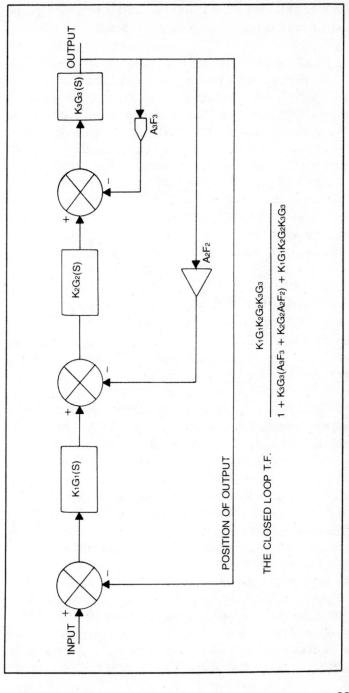

$$\frac{K_1 G_1 K_2 G_2 K_3 G_3}{1 + K_3 G_3 (A_3 F_3 + K_2 G_2 A_2 F_2) + K_1 K_1 K_2 G_2 K_3 G_3}$$

THE CLOSED LOOP T.F.

Fig. 6-17. A multiloop servosystem suitable for robotics applications.

229

then look at it, and put it in a computer for solution and see what we have done! Sometimes that is amazing and surprising!

If our system tends to hunt or oscillate about a final position then we might have to increase the friction or damping term of the system to effect a cure. We might do this by as simple a technique as adjusting a potentiometer. If we find the robot arm moves too slowly into a final position, so that a part goes by before the arm grippers can grasp it, then we might adjust the K term—also by adjusting a potentiometer—and correct the delay. We might find we have to adjust many potentiometers or controls all at the same time for a particular installation to make the robot work exactly as it should in that application. There can, and will be, in the normal course of events, many feedbacks in a robotics system and all these are necessary.

ELIMINATING THE ERROR

A closed loop servosystem operates on an error signal. The open loop systems operate on a command input signal. The closed loop systems always compare the output to the input and the difference between them is the error signal, it seems we always must have an error signal for the system to operate. We have discussed some problems associated with trying to minimize that error. It is possible to eliminate the operating error if one uses integral control methods!

What this kind of control does, in effect, is to average up the amount of signal developed in an error and then, through its block of control, develop a voltage proportional to that error and of the proper polarity to cause the motor controller of the robotics system to eliminate the error entirely! A very simplified block diagram of such a system is illustrated in Fig. 6-17.

Here we see the integration amplifier in the forward loop of the system. It is possible to obtain this effect with a proper compensation in the feedback loop also, but we do not obtain complete elimination of the error so easily. In effect we make the system very sensitive and so it can easily become unstable if there is no other compensation used to damp out any

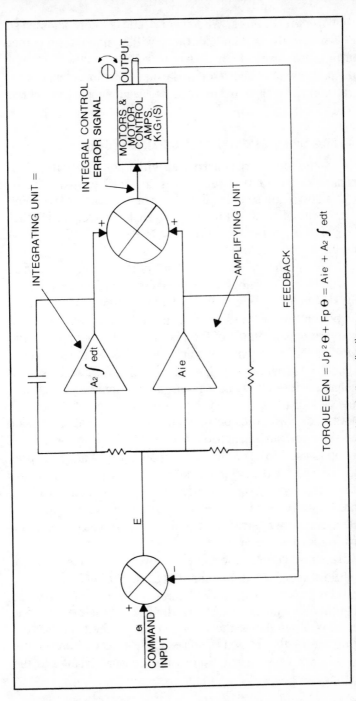

Fig. 6-18. An integral-control servosystem may be used in robotic applications.

231

oscillatory tendencies. When everything is finely adjusted so it does operate as it should, there will be no error in output but if there is any slight change to voltages in the system it might throw the balance off and cause some problems. Integral control is a delicate type of servomechanism, but it exists.

DETAILS OF ELECTROHYDRAULIC SERVOSYSTEM

Some robots are constructed with a combination of electrically operated transfer valves and either hydraulic or pneumatic pistons controlling the arm movements. Examine a cut-away of an actual hydraulic transfer valve. Notice that the valve is operated by torque motors. Figure 6-19 shows the cut-away illustration.

Locate the valve spool. This is the element which is moved by the torque motor to permit oil to flow to a piston, not shown here. This valve controls the input to one side of the piston while at the same time permitting the exhaust of oil from the opposite side of the piston back to a sump of some type. In order to have a quick response to the electrical signals, it is customary to have the spool dither, which simply means it vibrates in accord with a low amplitude, high frequency signal imposed on the control winding of the torque motor along with the control signals. Realize that it is easier to keep something in motion than to start it in motion, so this action is used to keep the valve spool moving and thus prevent delays due to sticking deadband. Deadband is that region of no response where a change in the control signal will *not* cause a corresponding pressure change of the piston.

Some other terms used to describe phases of operation of a transfer valve are:

null—the condition where the servovalve supplies zero control flow at balanced load forces across the piston.

null shift—the condition which results from a change in the bias current which produces a null. The shift is expressed in terms of a percent of the rated current of the servovalve.

deadband—that region of no response where a control signal will not cause a corresponding pressure change in the piston.

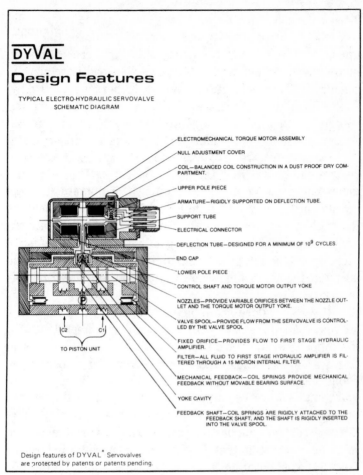

Fig. 6-19. Electrohydraulic servovalve-schematic diagram (courtesy Dynamic Valve Co.).

linearity—the degree to which the normal flow curve conforms to the normal flow gain line with the other operational variables held constant. It is measured as the maximum deviation of the normal flow curve from the normal flow gain line and is expressed as a percent of rated current.

threshold—the increment required to change from an increasing output to a decreasing output.

Since we have mentioned curves, let us examine the curve of normal input and output control tolerance for the type of servovalve under discussion. Look at Fig. 6-19.

Notice the tolerance lines, dotted on each side of the solid reference line. By taking the span between the tolerance lines, in the vertical direction (y axis), one can obtain the increment by which a percentage of the load flow may be calculated. The illustration shows a 10 percent rated flow. Also a projection on the X axis will give the percent of rated current. Both English and Metric units can be used to describe flow and other important dimensions as illustrated.

Finally, we want to examine two actual, commercial servodriver units from Dynamic Valve Co. These are the electronic parts of a servosystem which operate the servovalve previously shown. In Fig. 6-21A we see a schematic for a velocity servodriver, and we note that it has an integral control element in the forward gain loop. There is a control there called the integrator sensitivity control and this can be adjusted to obtain the desired effect on the servo performance.

If we consider for a moment the effect of an error in a servosystem in which velocity of movement is important, and is an input specified quantity, we will realize that an error in this quantity will cause the wrong velocity, or an oscillation about some normal desired value. We don't want such an error in this type of system. We want it to get to the desired and commanded velocity and hold that velocity. The integral-control type of servo can make this possible. Notice in that same system that a tachometer is used for derivative feedback compensation. This produces the stability required.

Sometimes it is desirable to be able to open the loop of a closed loop servosystem in order to check signals at various points and in some cases to locate troubles. In a closed loop system, where the system always tries to minimize its error, one might have a lot of trouble trying to see what kinds of signals are actually being produced by the various blocks of the system. When we open the loop we can create a static condition which permits us to see what these signals are, since there is no way the system can function to reduce the error as we have previously defined the error signal. In block A of Fig. 6-21 we see the switch which opens the loop for testing or adjustment or whatever.

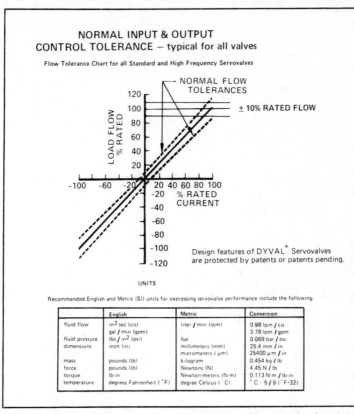

Fig. 6-20. Servovalve curve showing characteristics of its operation (courtesy Dynamic Valve Co.).

In block B of Fig. 6-20 we examine a positioning type of system. There is a control to adjust the value of the gain (K), and there is compensation, with derivative feedback of an electronic type. In a way it is like the tachometer feedback, but doesn't operate from the moving output element as the tachometer type does. This type operates from the signal and is able to determine changes in the signal and the rate of change of the command or error signal. There is also a velocity feedback to the input and various other controls for balance and monitoring purposes. In all cases the output of the servodriver units is sent to the torque motor on the hydraulic transfer-valve which, in turn, controls the output piston producing the power and movement required of the system.

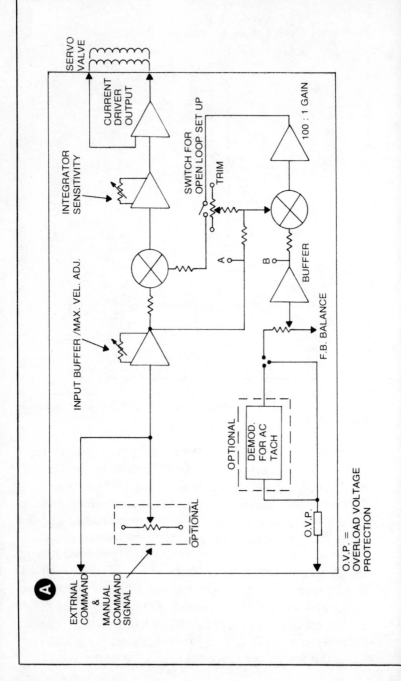

A

SERVO VALVE

CURRENT DRIVER OUTPUT

100 : 1 GAIN

INTEGRATOR SENSITIVITY

SWITCH FOR OPEN LOOP SET UP

TRIM

A   B

BUFFER

INPUT BUFFER / MAX. VEL. ADJ.

F.B. BALANCE

OPTIONAL

DEMOD. FOR AC TACH

OPTIONAL

O.V.P.

EXTRNAL COMMAND & MANUAL COMMAND SIGNAL

O.V.P. = OVERLOAD VOLTAGE PROTECTION

236

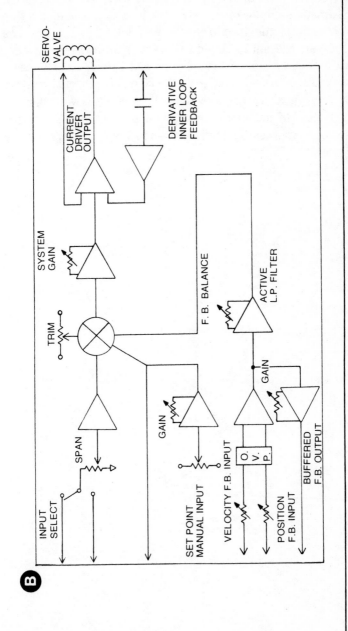

Fig. 6-21. Servo Driver unit electronics schematics: (A) An integral control **velocity driver amplifier**. (B) A positioning servodriver amplifier.

237

We should not conclude this chapter without mentioning that Dyval, Inc. also makes a transfer valve which has a pneumatic element mounted on the hydraulic valve, and this pneumatic element controls the hydraulic flow. In systems where air pressure is used to control fluid flow, the system has a combination of both power and speed. Recall that a pneumatic system is said to be the fastest mechanical system because the air can be moved faster than oil, however, air is compressible and so for power, without elasticity, one needs the non-compressible hydraulic fluid as the power producing element.

A Computer-
Operated Hobby Robot

The robotic arm described in the following pages, courtesy of Microbot and *Robotics Age* magazine is one which its inventors say can be used for hobby purposes, for research purposes in computer program development, or just for robotics experimentation. The arm is operated by a computer as illustrated in Fig. 7-1.

In the operation shown, the arm is being directed to pick-up and stack some small wooden blocks in a sequential order. This is not a professional application, but is fun, as is the writing of the computer program to accomplish such tasks. You can see the relative size of the arm. You might remember GARCAN and wonder if such an arm might not be attached to that paraplegic robot. One advantage of this unit's size is that you can develop various computer languages and programs for robotic use, and then check these programs without having to resort to the expense of procuring a full size robot.

Many universities and research centers are now developing the necessary communication languages for robot control and programming, using computerized programs which are of great complexity, yet plan their own manipulation strategies, efficiently performing complex coordinate

239

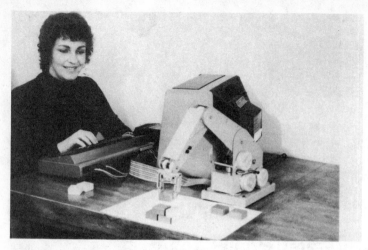

Fig. 7-1. Computer operation of the hobby-research robot arm is easy (courtesy Microbot).

transformation mathematics. Studies are being made to use more and better sensors and get better use from the available types of sensors, and studies are being constantly made to gain a better understanding of the robot's kinematics. The Mini-Mover 5 as this robot arm has been named, was designed to assist this kind of research.

LOOKING AT THE ARM DESIGN AND OPERATION

In Fig. 7-2 we see a close-up of the Mini-Mover 5 with its gripper and three of its 6 stepping motors. These control the arm and gripper movement. The motors are; base drive, shoulder drive, elbow drive, 2 for wrist drive, and the gripper. In the upper right hand corner of the illustration we can see a portion of the computer used to control this unit.

Because a coordinate system must be designated for the arm and the gripper—and this is true for any robot system—we show in Fig. 7-3 how this is designated for the Mini-Mover 5. You will notice that a rectangular system of coordinates is used which has its origin at that point where the base of the arm pivots on a supporting surface. The gripper, then, will be displaced from the origin in some X, Y coordinate direction, and have an altitude position in the Z

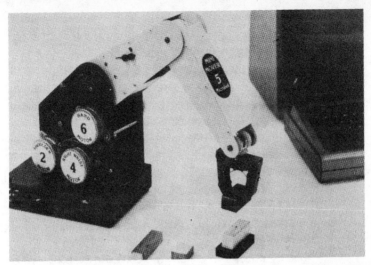

Fig. 7-2. A close view of the Mini-Mover 5, as the arm is called, and three of the six motors used in its operation. Motors 1, 3, and 5 are on the opposite side of the arm in this view (courtesy Microbot).

direction. One might wonder if using a system of polar coordinates wouldn't identify the gripper position much easier?

Remember that the arm will be responding to a series of pulses which, in turn, will step the various stepping motors.

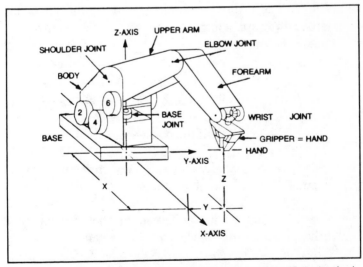

Fig. 7-3. The coordinate system for the robotic arm (courtesy Robotics Age).

Thus it might be easier to program the unit to move five units right (X) and two units out (Y) and drop the gripper itself down two units (Z) to engage some object.

We note that this arm's gripper has five degrees of freedom because it is driven by a differential gear mechanism which can rotate the gripper in both pitch and roll. Add this to the motions of the body, shoulder, elbow, and wrist and you have a sufficient number of degrees of freedom so that the arm can be accurately positioned and the gripper positioned in a partial sphere whose radius is 17.5 inches from the origin.

We have mentioned earlier in this book that in some cases of robot design, one might use motors which drive various elements through the use of cables or linkages. That is the case with this robot arm, as shown in Fig. 7-4.

The drive motor for each joint consists of a stepping motor, reduction gearing, and a cable drum. From each drum a tensioned cable goes out over pulleys to the member being driven and then may or may not return to the drum. Rotation of this drum causes rotation of each member in proportion to the diameter of the drive pulley attached to that member and the drum's diameter. The movement of one joint on this arm may result, therefore, in the motion of another joint, and this may not be desirable. Here is how the designers of the Mini-Mover 5 describe the cabling operation, which must be understood if you are going to compensate for these unwanted movements by properly programming the operation of the arm:

"*Base Rotation* (Joint 1)—The base drive cable passes over two idler pulleys, making a 90 degree bend, to a drive pulley fixed to the base. The base drive motor causes the entire arm to rotate about the base joint.

"*Shoulder Bend* (Joint 2)—The shoulder drive cable passes around the drive pulley on the upper arm segment to rotate it about the shoulder joint.

"*Elbow Bend* (Joint 3)—The elbow drive cable passes around an idler pulley on the shoulder axis to a drive pulley fixed to the lower arm segment. Joints 2 and 3 interact, so that changing the shoulder angle results in an equal change in the elbow angle. This interaction has

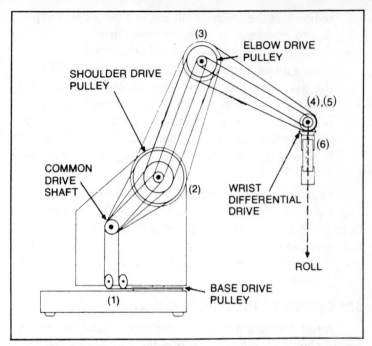

Fig. 7-4. The cable diagram for the Mini-Mover 5 robotic arm (courtesy Robotics Age).

been designed so that the elbow drive, in effect, controls the angle of the forearm with respect to the horizontal. However, the limits of forearm motion are measured with respect to the upper arm, not the horizontal (x-y) plane.

"Wrist (Joint 4, left and Joint 5 right)—The fourth and fifth drive cables pass around idler pulleys on both the shoulder and elbow joints and terminate on the drive pulleys of the left and right wrist differential gears. The interaction of these joints with joints 2 and 3 has the effect that the wrist drives control the angle of the hand with respect to the horizontal (pitch) and the rotation of the hand around its pointing vector (roll). Through the action of the differential, the pitch angle P equals the average of the positions of the left and right gears and the roll angle R is their difference. The range of motion is limited to ±270 degrees at each of the wrist gears due

to limits of the cable length and to ±90 degrees in pitch due to interference from the lower arm.

"*Hand* (Joint 6)—The hand drive cable passes over idler pulleys located on the shoulder and elbow joints, through the center of the wrist differential, and finally terminates at the hand. This drive interacts with joint 3 (elbow) in that elbow bend will cause the hand to open slightly. This can be compensated for by closing the hand exactly the same number of steps that the elbow is raised."

Examine a view of the hand and its differential gearing in Fig. 7-5. It is interesting to examine how the grippers have been made for this robot. Each section is like a two-section finger, and in the finger tip portion are surfaces which permit gripping an object so it won't slide away.

SOME NOTES ON THE GRIPPER OPERATION

Another diagram of the cabling arrangement for the gripper is shown in Fig. 7-7. Notice the torsion spring which permits the hand to open when the cable tension slackens. Also notice the switch which could prevent a too-tight grip and thus causes cable breakage. This also has a second function, which is to cause the cable to tighten slightly more after it first grasps an object. This switch sends a signal to the computer that an object has been grasped and so, if desired, one might program the computer so that when it gets this signal, it can cause another step or so of the tension motor so that a good tight grip results.

OPERATING THE STEPPING MOTORS

Each of the motors used has two windings which are center tapped. A transistor then can operate on each half winding in such a way that a rotating magnetic field is established which then can cause the motor armature to step around in small steps. These motors require twelve volts which is supplied through transistors, driven by the controller, as shown in Fig. 7-8. The controlling computer can, with

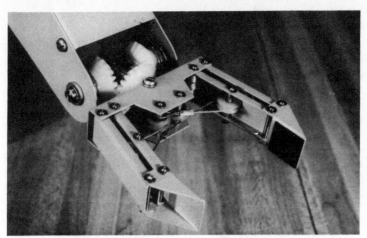

Fig. 7-5. The Mini-Mover 5 hand (courtesy of Microbot).

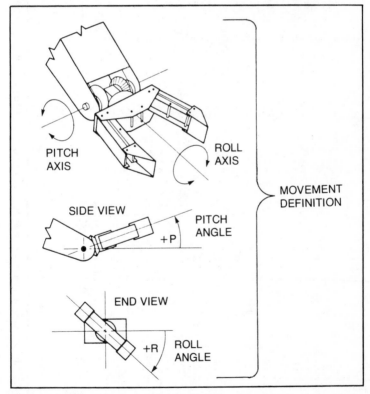

Fig. 7-6. Mini-Mover 5 hand gearing and cable arrangement to grippers (courtesy Microbot).

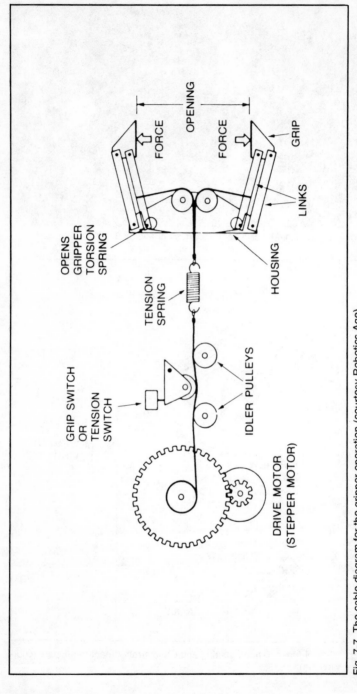

Fig. 7-7. The cable diagram for the gripper operation (courtesy Robotics Age).

246

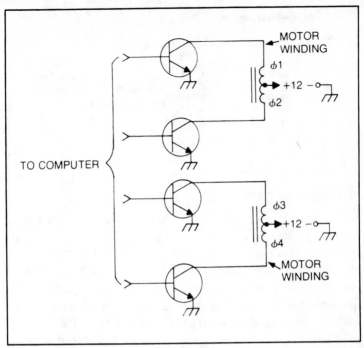

Fig. 7-8. The transistor drivers for two stepper motors. Since each arm element has a "forward and reverse" direction, when you step one motor, you must also step the other to permit movement (courtesy Microbot).

proper software, apply almost any desired arrangement of currents to these windings.

THE COMPUTER-CONTROL OUTPUT

If you note the designation of the motor windings you can see, in Table 7-1, the outputs possible to make the stepper motor move either forward or backward. The designers tell us that to step the motor clockwise the patterns are output sequentially from the top of the table to the bottom of the table and when the end of this table is reached, it is necessary to wrap around to the other end of the table and continue the outputs sequentially. To reverse the motor rotation, the operation is reversed. With this table the motor can be made to step through one full revolution when it has received 96 pulses which makes the increments small enough so that a smooth operation results.

TO WINDINGS DRIVERS			
$\phi1$	$\phi2$	$\phi3$	$\phi4$
0	1	0	1
0	1	0	0
0	1	1	0
0	0	1	0
1	0	1	0
1	0	0	0
1	0	0	1
0	0	0	1

(left side: CLOCKWISE, downward arrow; right side: COUNTERCLOCKWISE, upward arrow)

Table 7-1. The Computer Code to Rotate the Stepper Motors of Fig. 7-8 (courtesy Microbot).

THE COMPUTER INTERFACE

The computer interface is shown in Fig. 7-9 and it is so designed that it will operate, by changing some internal jumpers, with any 8 bit parallel I/O port (TTL levels). The interface was operated by a Radio Shack TRS 80 Level II computer by the designers of the robot arm. Notice that there are seven 4-bit parallel outputs and a single 4-bit input. The address information is used to channel the four bits of output data to the particular motor drive circuit selected. This is done through individual 4-bit latch circuits. After a phase rotation pattern has been sent to one motor through a given latch, that latch is held until the next phase pattern is transmitted. These latch circuits actually control the transistor drivers, and thus the power which is applied to the motor coils to make the motor step.

As we examine Fig. 7-9 notice several facts. There is one input to the data bus from the arm at port A. There is one spare output (port B) since only six of the output ports are needed for control of the six stepper motors used on the arm. This output port might be used to control another motor if desired. Also, notice that only one feedback line is used. This is to return a signal from the tension or grip-switch of Fig. 7-7. You could have three more feedbacks incorporated in this system if desired, without increasing the size of the I/O port.

THE ARMBASIC COMMAND COMPUTER LANGUAGE

The ArmBasic language, developed to control this robot arm, is an extension of the BASIC language for the TRS-80 computer and consists of the five additional commands necessary for complete control of the arm. The commands chosen are:

- @STEP
- @CLOSE

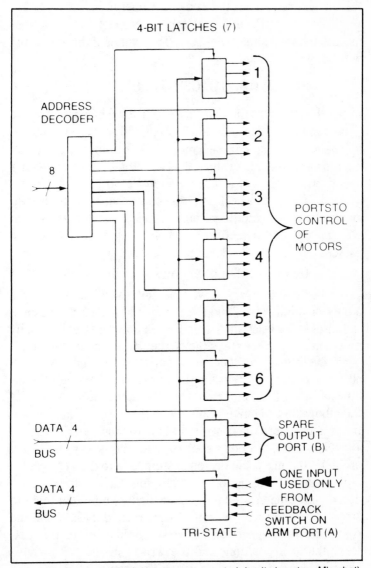

Fig. 7-9. The computer interface to the motor control circuits (courtesy Microbot).

- @SET
- @RESET
- @READ

Each of these commands is used as a BASIC statement, and interpreted as a Z-80 machine code subroutine. Here is how the designers explain the computer operation.

"The @ STEP command causes each of the 6 stepper motors to move simultaneously. The syntax of this command will be:

@ STEP (D), (J1), (J2), (J3), (J4), (J5), (J6)

in which statement the D is an expression which governs the delay between steps and the J1 through J6 are the statements for the number of steps each motor is to be moved. The delay expression, which is evaluated as an integer, determines the speed of stepping. If a delay value of zero is indicated the system is set to wait 1.2 milliseconds between each step signal. For each additional unit of delay, the driver system will wait an additional .03 milliseconds between the stepping pulses. Therefore the delay is expressed as:

Delay = 1.2 + 0.03 times D milliseconds

"To determine which way the motors will step, one uses a plus or minus sign, and the number of steps; the value of the integer for each motor, indicates how far it will actually move, i.e., the more steps the farther it will move the arm."

One interesting concept of this system is that there is an automatic compensation for the interaction between the elbow joint and the hand openings. The compensation permits the elbow joint to move while not affecting the hand opening. In systems where the control motors are mounted right at the joints and are operated by specific signals this may never be a problem, but in a cable driven system, due to the very nature of the cabling, interaction is something that has to be considered and compensated for. The designers point out that the steps J4 and J5 refer to motors which drive the left and the right differential gears of the wrist, so that the relation between the gear positions and the wrist roll and pitch angles must be determined by the user's program.

An important function performed by the @ STEP command is to coordinate unequal motor commands linearly. Thus if the base control motor is commanded to move X number of steps and the shoulder motor is commanded to move Y number of steps where Y<X there should be in such a system an appropriate number of delays in the Y stepping signal generation so that the timing of the Y steps will correspond, in time, to the X number of steps. If this is so then the arm movement is smooth and flowing, just as your arm movement would be as you reach for a pencil. When the number of steps for each section are accomplished in the same time frame, then the end point segments of the arm will reach their final positions simultaneously.

The @ CLOSE command which is the command which causes the hand (or grippers) to close until the grip-switch is activated and sends back a signal to the computer that the hand has closed with the designed pressure effective. The syntax for the gripper closing is also very simple:

@ CLOSE (D)

And the letter D again represents the delay-in-closing time.

We next consider the @ SET command which has a syntax like this:

@ SET (D)

This is a very simple syntax arrangement indeed. D is actually the system delay that governs the speed.

The designers state that when one uses the @ SET command it puts the control system into a manual mode so that when you depress the various keys on the TRS-80 keyboard (or keyboard of an equivalent computing system) each joint of the arm can be positioned individually. One keyboard arrangement for controlling the arm is shown in Fig. 7-10.

When you wish to stop this manual mode of operation, you simply depress the zero key. Since the driver for the output is capable of reading all keys at one time, if you hold down more than one key, the arm will move in accord with the combination of signals, i.e., the joints commanded will move at the same time. The motion will continue as long as the keys

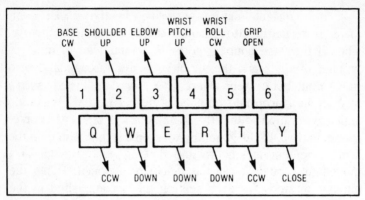

Fig. 7-10. One keyboard control system for the robotic arm Mini-Mover 5 (courtesy Microbot).

are held down. To stop the motion one simply releases the keys.

By now you can imagine sitting down at a computer keyboard and carefully depressing the appropriate keys to cause a Microbot arm to move in the manner you desire. By carefully choosing your key and the timing of its depression, you can make the arm move over a block, extend down and grip it, move back upward so it can turn with its load and move it to a new location and carefully stack the blocks as shown in Fig. 7-1.

It is important to note that as each command is executed, the number of steps commanded for each motor in the arm are counted and stored in a set of registers in RAM. On the video display one can determine the commanded position of the arm under either manual or computer control. There are two commands which access these registers. They are the following:

- @ RESET
- @ READ

The @ RESET command zeros the position registers and stops all current to the motors, thus allowing you to physically move the arm to a new position. Any other command cancels @ RESET, and the position of the arm at that time becomes the new home position.

The syntax of the @READ command is:

@READ (V1), (V2), (V3), (V4), (V5), (V6)

Where V1 to V6 are any BASIC variables. This command reads the six position registers and transfers their contents into the variables. In this way the current position of the robotic arm is available to the BASIC program at any time.

PROGRAMMING THE MINI-MOVER 5 ROBOTIC ARM

Dr. John Hill, Design Engineer of Microbot, explained the way to program this robotic arm in the Spring, 1980 issue of the *Robotics Age* magazine.

"To illustrate the use of ARMBASIC in manipulation, we will discuss several methods of programming a pick-and-place task. The task consists of repeatedly picking objects up at one place and setting them down in another. This task is common in industry where parts from a feeder are deposited in another machine or onto a moving conveyor, for instance. To simplify the example, we will assume an unending supply of parts at the pick-up point and continual removal of parts at the destination point.

The task is defined in Fig. 7-10. Pickup is from A and placement is at D. The approach and departure trajectories, AB and CD are usually required for practical reasons: an object resting on a flat surface normally must be lifted before it is translated instead of just sliding it along the surface. In an industrial environment, such a movement could possibly damage the part, the manipulator, or some other object in the workspace.

"A simple sequence of ARMBASIC commands for performing the pick-and-place task is shown in Table 7-2. Movements are performed by the @STEP command, including opening the gripper (statement 60), $P=(P1, P2, P3, P4, P5)$, $Q=(Q1, Q2, Q3, Q4, Q5)$ and $R=(R1, R2, R3, R4, R5)$ define vectors whose elements are the number of steps of each joint needed to move the arm from A to B, B to C and C to D respectively. S is the

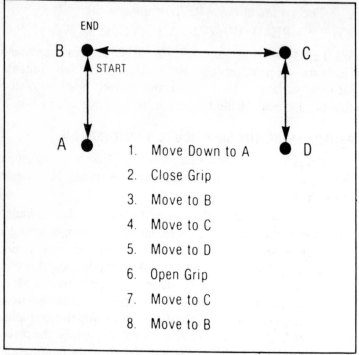

Fig. 7-11. The robot's arm movements in a Pick-and-Place program of Table 7-2 (courtesy Robotics Age).

number specifying the speed and GR is the number of steps the gripper is to be opened to release the part.

"This simple approach is not very practical, as it requires that the number of steps on each joint be known in advance. One practical approach for small tasks is entering the data manually by moving the arm to each of the four positions (A, B, C and D of Fig. 7-11) and letting the computer count and remember the number of steps each joint moves. This operation is frequently called "teach mode."

The teach program shown in Table 7-3 illustrates this principal. Operating interactively with the computer, the operator manually positions the arm to each of the four positions to specify the pick-and-place task. Each position is obtained by moving the arm, joint-by-joint, using the keys in the upper left hand corner of the keyboard as previously described under the @SET command. Even the desired grip opening is conveyed by teach control. Only the speed is entered numerically.

"After the arm is positioned at each of the four locations, the values of the software position counters are read by the @READ command and stored in the vectors A, B, C and D. The elements of these vectors, A=(A1, A2, A3, A4, A5), B=(B1, B2, B3, B4, B5), C=(C1, C2, C3, C4, C5), and D=(D1, D2, D3, D4, D5), are the values of the position counters of each joint. The P, Q, and R vectors of the simple program (Table 7-3) can therefore, be obtained by subtraction as follows: P=B−A, Q=C−B, R=D−C. The desired grip opening, GR, is measured by reading the grip position counter before and after closing the grip using the @CLOSE command. (Statements 170-200 in Table 7-2.)

Examine Table 7-4. For this program, the locations of the four positions of the pick-and-place task (A, B, C, D) are defined by their Cartesian coordinates (see Fig. 7-3). The x, y, z, coordinates (in inches) and pitch and roll angles (in degrees) at the fingertips are:

$$A=(8, 0, 0.5, -90, 0)$$
$$B=(8, 0, 1.5, -90, 0)$$
$$C=(6, 5, 1.5, -90, 90)$$
$$D=(6, 5, 0.5, -90, 90)$$

Table 7-2. Simple Implementation of ARMBASIC Pick-and-Place Program (Courtesy Robotics Age).

10	@ STEP	S,	−P1,	−P2,	−P3,	−P4,	−P5	
20	@ CLOSE							
30	@ STEP	S,	P1,	P2,	P3,	P4,	P5	
40	@ STEP	S,	Q1,	Q2,	Q3,	Q4,	Q5	
50	@ STEP	S,	R1,	R2,	R3,	R4,	R5	
60	@ STEP	S,	0,	0,	0,	0,	0,	GR
70	@ STEP	S,	−R1,	−R2,	−R3,	−R4,	−R5	
80	@ STEP	S,	−Q1,	−Q2,	−Q3,	−Q4,	−Q5	
90	GOTO	10						

Note that the object is gripped at a point 0.5 inches above the table top (z=0) and is raised (or lowered) 1.0 inches for lift off (set down). Its x location is changed from 8 to 6 inches and its y location changed from 0 to 5 inches. The angular orientation of the gripper is always straight down (pitch=−90 degrees), but the object is rotated 90 degrees on its vertical axis between pickup and set down (roll changes from 0 to 90 degrees).

"Data for the Cartesian coordinate program is located after statement 10000. Each data statement specifies an arm position and orientation in terms of x, y, z, pitch and roll, and also includes a sixth parameter, governing the grip opening (if positive) or squeezing force (if negative) after each move, and a seventh parameter governing the speed setting during each move. The actual arm solution, beginning at statement 5000, takes an arm configuration specified by the variables X, Y, Z, P, and R, and finds the joint angles (in radians) that would place the arm in that configuration. It also includes several tests to determine if the position and orientation requested is within the reach of the arm.

"The first data statement gives the initialization point, which is used by the initialization phase of the program, statements 140-260, to tell the computer where the arm is in the coordinate system. Note that the arm position registers are not used by the program (although the @ RESET is necessary). Instead, the operator is instructed to position the arm at the initialization point under control of the @SET command. The joint positions corresponding to that point are computed and stored in variables W1-W5. In the main part of the program, the relative offsets from one point to the next are computed and used to step the arm. Thus, all arm movements are made relative to the initialization point.

"Since the computer has no way of measuring the arm's joint angles upon startup, such an initialization must be made by every program that relies on an external coordinate system. Similarly, all arm movements are made "open loop," or without position feedback. There-

Table 7-3. The Mini-Mover 5 Teaching Program (Courtesy Microbot).

```
10 REM
20 REM
30 REM
40 REM
50 REM
60 REM
70 REM
80 REM
90 REM
100 @RESET: REM        ZERO COUNTERS
110 PRINT "PICK - AND - PLACE ROUTINE"
120 PRINT "    USE MANUAL KEYS TO POSITION ARM"
130 INPUT "SPEED ="; S
140 PRINT "POSITION GRIPPER ON PART, TYPE 0 WHEN DONE"
150 PRINT "    ADJUST GRIP OPENING TO CLEAR PART"
160 @SET
170 @READ A1,A2,A3,A4,A5,C
180 @CLOSE:  REM    CLOSE GRIPPER AND MEASURE PART
190 @READ A1,A2,A3,A4,A5,GO
200 G = GO - GC :REM    GRIP SIZE OPEN LESS GRIP SIZE CLOSED
210 PRINT "POSITION PART ABOVE PICKUP SITE, TYPE 0 WHEN DONE"
220 @SET
230 @READ B1,B2,B3,B4,B5
240 PRINT "POSITION PART ABOVE PLACEMENT SITE, TYPE 0 WHEN DONE"
250 @SET
260 @READ C1,C2,C3,C4,C5
270 PRINT "POSITION PART AT PLACEMENT SITE, TYPE 0 WHEN DONE"
280 @SET
290 @READ D1,D2,D3,D4,D5
300 REM
310 REM    RELEASE PART AND RETURN TO B
320 REM
330 @STEP S,0,0,0,0,0,G
340 @STEP S, C1-D1, C2-D2, C3-D3, C4-D4, C5-D5
350 @STEP S, B1-C1, B2-C2, B3-C3, B4-C4, B5-C5
360 REM
370 REM    WAIT UNTIL READY
380 REM
390 INPUT "TYPE G TO GO"; R$
400 IF R$ = "G" THEN 410 ELSE 390
410 REM
420 REM    RUN THE PICK - AND - PLACE PROGRAM
430 REM
1000 @STEP S, A1-B1, A2-B2, A3-B3, A4-B4, A5-B5
1010 @CLOSE
1020 @STEP S, B1-A1, B2-A2, B3-A3, B4-A4, B5-A5
1030 @STEP S, C1-B1, C2-B2, C3-B3, C4-B4, C5-B5
1040 @STEP S, D1-C1, D2-C2, D3-C3, D4-C4, D5-C5
1050 @STEP S,0,0,0,0,0,G
1060 @STEP S, C1-D1, C2-D2, C3-D3, C4-D4, C5-D5
1070 @STEP S, B1-C1, B2-C2, B3-C3, B4-C4, B5-C5
1080 GOTO 1000
1090 END
```

257

Table 7-4. A Cartesian Coordinate Program for Mini-Mover 5 (Courtesy Microbot).

```
10 REM        ******************************************************
20 REM        *                                                    *
30 REM        *                 MINI MOVER 5                       *
40 REM        *                                                    *
50 REM        *          CARTESIAN COORDINATE CONTROL              *
60 REM        *                                                    *
70 REM        *                   PROGRAM                          *
80 REM        *                                                    *
90 REM        ******************************************************
100 REM     DEFINE ARM CONSTANTS
101 H=8.1     :REM   SHOULDER HEIGHT ABOVE TABLE
102 L=7.0     :REM   SHOULDER TO ELBOW AND ELBOW TO WRIST LENGTH
103 LL=3.5    :REM   WRIST TO FINGERTIP LENGTH
104 REM
110 REM        DEFINE OTHER CONSTANTS
111 PI=3.14159
112 C=180/PI         :REM   DEGREES IN 1.00 RADIAN
113 R1=1             :REM   FLAG FOR WORLD COORDINATES
114 REM
120 REM        DEFINE ARM SCALE FACTORS
121 S1=937.8: S2=S1  :REM   STEPS/RADIAN, JOINTS 1 & 2
122 S3=551.4         :REM   STEPS/RADIAN, JOINT 3
123 S4=203.7: S5=S4  :REM   STEPS/RADIAN, JOINTS 4 & 5
124 S6=618           :REM   STEPS/INCH, HAND
125 REM
130 REM          INITIALIZATION
131 DIM UU(7,50)     :REM ROOM FOR 50 MOVES
132 U=0
133 @RESET
134 REM
140 REM              READ IN FIRST DATA LINE FOR INITIALIZATION
141 READ X,Y,Z,P,R,GR,S
150 PRINT "SET ARM TO THE FOLLOWING POSITION & ORIENTATION"
151 PRINT "   USING KEYBOARD, TYPE O WHEN FINISHED"
152 PRINT "      X = "; X; "INCHES"
153 PRINT "      Y = "; Y; "INCHES"
154 PRINT "      Z = "; Z; "INCHES"
155 PRINT "   PITCH = "; P; "DEGREES"
156 PRINT "    ROLL = "; R; "DEGREES"
157 PRINT "    HAND = "; GR/S6; "INCHES"
170 @SET
200 GOSUB 5000       :REM GET JOINT ANGLES FOR INITIALIZATION
210 REM  W IS WHERE WE ARE AT
220 W1=INT(S1*T1)
230 W2=INT(S2*T2)
240 W3=INT(S3*T3)
250 W4=INT(S4*T4)
260 W5=INT(S5*T5)
270 REM
300 REM READ IN NEXT POINT
308 READ X,Y,Z,P,R,GR,S
309 IF X < -100 GOTO 1000
310 PRINT "MOVE TO X,Y,Z,P,R,GR,S"
311 PRINT X,Y,Z,P,R,GR,S
320 GOSUB 5000       :REM GET JOINT ANGLES FOR NEXT POINT
325 REM        C IS THE CHANGE (IN COUNTS)
330 C1=INT(S1*T1)-W1
340 C2=INT(S2*T2)-W2
350 C3=INT(S3*T3)-W3
360 C4=INT(S4*T4)-W4
370 C5=INT(S5*T5)-W5
400 REM   UPDATE WHERE WE WILL BE
410 W1=W1+C1:W2=W2+C2:W3=W3+C3:W4=W4+C4:W5=W5+C5
420 @STEP S,C1,-C2,C3,-C4,C5
425 U=U+1
430 UU(1,U)=C1
431 UU(2,U)=-C2
```

258

```
 432 UU(3,U)=C3
 433 UU(4,U)=-C4
 434 UU(5,U)=C5
 450 UU(6,U)=GR
 455 UU(7,U)=S
 460 IF GR <0 GOTO 500
 461 REM    OPEN GRIP IF GR>0
 470 @STEP S,0,0,0,0,0,GR
 480 GOTO 300
 490 REM    CLOSE GRIP AND SQUEEZE IF GR < 0
 500 @CLOSE
 520 @STEP 100,0,0,0,0,0,GR
 530 GOTO 300
 540 REM
1000 PRINT " RUN THE PROGRAM"
1010 FOR I=2 TO U
1020 @STEP UU(7,I),UU(1,I),UU(2,I),UU(3,I),UU(4,I),UU(5,I)
1030 IF UU(6,I)<0 GOTO 1060
1040 @STEP UU(7,I),0,0,0,0,0,UU(6,I)
1050 GOTO 1100
1060 @CLOSE
1080 @STEP 100,0,0,0,0,0,UU(6,I)
1100 NEXT I
1110 GOTO 1010
2000 END
2010 REM
5000 REM        SUBROUTINE TO CALCULATE JOINT COORDINATES
5010 REM           ENTER WITH X,Y,Z,PITCH,ROLL, AND R1
5015 REM           RETURN WITH JOINT ANGLES T1 TO T5
                                        (IN RADIANS)
5020 P=P/C:R=R/C
5030 RR=SQR(X*X+Y*Y)
5040 IF RR<2.25 THEN 6000
5050 IF X=0 THEN T1=SGN(Y)*PI/2 ELSE T1=ATN(Y/X)
5060 IF X<0 GOTO 6000
5070 T4=P+R+R1*T1
5080 T5=P-R-R1*T1
5090 IF T4<-270/C OR T4>270/C GOTO 6000
5100 IF T5<-270/C OR T5>270/C GOTO 6000
5110 RO=RR-LL*COS(P)
5120 ZO=Z-LL*SIN(P)-H
5130 IF RO<2.25 GOTO 6000
5140 IF RO=0 THEN G=SGN(ZO)*PI/2 ELSE G=ATN(ZO/RO)
5150 A=RO*RO + ZO*ZO
5160 A=4*L*L/A-1
5170 IF A<0 GOTO 6000
5180 A=ATN(SQR(A))
5190 T2=A+G
5200 T3=G-A
5210 IF T2>144/C OR T2<-127/C GOTO 6000
5220 IF (T2-T3)<0 OR (T2-T3)>149/C GOTO 6000
5230 RETURN
6000 PRINT "********* OUT OF REACH *********"
6010 END
6020 REM
9000 REM    COORDINATE DATA FOR PICK-AND-PLACE TASK
9010 REM        STATEMENT 10000 IS INITIALIZATION POINT
9020 REM        STATEMENT 10010-10070 DEFINE TRAJECTORY
9030 REM        STATEMENT 10080 SIGNALS NO MORE DATA
10000 DATA     5,  0,   0,  -90,   0,   0,  50
10010 DATA     8,  0, 1.5,  -90,   0,   0,  50
10020 DATA     8,  0, 0.5,  -90,   0, -30,  50
10030 DATA     8,  0, 1.5,  -90,   0,   0, 100
10040 DATA     6,  5, 1.5,  -90,  90,   0,  50
10050 DATA     6,  5, 0.5,  -90,  90, 300,  50
10060 DATA     6,  5, 1.5,  -90,  90,   0,  50
10070 DATA     8,  0, 1.5,  -90,   0,   0,  50
10080 DATA  -999,  0,   0,    0,   0,   0,   0
```

259

fore, a reinitialization is necessary if the arm is unable to successfully perform a movement due to overloading of collision. The use of open loop control by stepper motors eliminates the need for joint position sensors and associated interface circuitry and is thus an important design tradeoff that reduces the cost of the arm while still providing accurate positioning.

"During the second phase of the program, statements 300-530, the arm is moved from data point to data point as the solutions are carried out. An arm solution takes less than one second. The joint angles corresponding to each data point are stored in array UU. After all the solutions are obtained, control is transferred to the third phase, beginning at statement 1000, where the pick-and-place cycle is run repeatedly without coordinate conversions by using the joint angles in UU.

"Those who analyze the Cartesian coordinate program carefully will note that it is, in fact, general purpose, capable of generating arbitrary motion trajectories with up to 50 movements. This is sufficient for many assembly tasks such as building structures from blocks. The power of the ARMBASIC approach can be seen by comparing the simplicity of this program, which is less than 100 lines of code, with previous approaches which have required 8K or more of assembly language coding to do the same type of task."

ROBOTIC-ARM ACCURACIES AND POWER

The Mini-Mover 5 was developed for, and can be used in learning and teaching, and in an experimental way to verify computer related concepts economically. From this viewpoint, it is an impressive device. From a useful standpoint of doing work it has some limitations which need examination. One of these limitations is the power capability of the arm.

There can be no doubt that in advanced robotics applications a significant amount of power must be generated, proportional to the size of the robot of course, to enable the machine to do useful work. One may argue that useful work is, indeed, a good reason for the machine's existence.

Another problem of machines of Mini-Mover 5 design may be the physical control cabling because it has, by virtue of its type, some play and elasticity which becomes more pronounced as it attacks jobs which require power and force for handling. This seems to defeat one much-desired goal of robot machine development; that of consistent and reproducible accuracy to an extremely high degree of precision in a repeatable process requiring movement. Cables give much flexibility in construction and in many cases simplify such construction. However, they may not produce the feedback effect of inertia in an automated system as much as gears do. Certainly, if one does not have to have motors and gears at each joint, the system inertia can be reduced for operational and computational analysis purposes. One might then avoid the complexities of having a variable moment of inertia to be concerned with when the robot arm is moved.

Currently the accuracy of robot arm device positioning is limited to some .004 inch in a repeatable processing cycle. While a higher degree of accuracy may be needed in some types of applications, we must admit that .004 inch is not bad!

COMMUNICATING WITH THE ROBOT

In the example of computer communication using the operation of Mini-Mover 5, hard-wire signal cabling was used. We do not have much to worry about as far as getting extraneous signals into the machine's operation, unless we are very careless and lay the signal cable in a strong electromagnetic field, improperly shield it, make the signal cable too long, or improperly terminate it at either end. Usually, since the signal cabling is relatively short in length, properly laid, terminated, and even may be shielded, there is no problem, and the signals are transmitted strongly and accurately.

If we have a robot located at some distance from the computer which directs its operations, and if the environment through which signal cables or lines to the robot must be laid is harsh, as far as signal transmission is concerned, then we could have a communications problem. The problem evolves around two concepts; the loss of signal information, and the production of spurious signal information.

Loss can be due to actual signal disintegration or to distortion to such an extent that the robot cannot tell that it is receiving a command or feedback signal. In an industrial environment two possibilities exist.

The robot with its drive motors is the end unit and the computer, servomechanism amplifiers, and such are all located at a remote point away from the robot. Thus the only signals going to the robot itself will be the voltages necessary to cause the motors to operate, and since these are high level signals they are less subject to interference. The second possibility is that the robot is almost entirely self contained and thus, with its integral computer and servo system, is relatively well protected from a harsh communications environment. In this latter case, the robot receives signals from its own world only, be they commands from the computer program, or feedback from its working sensors.

When the robot is in a fixed location, the wires connecting it to a remote computing and electronic control mechanism panel are usually contained in metallic tubing which gives physical protection and shielding if such tubing is connected to a common ground. Even then, a shielded cable may run inside this kind of conduit. In this type of application this usually takes care of the communications problem. In fact, in industrial plants, the use of hard-ware signal cables might be the most useful way to communicate with the robot, and have it communicate with its control base! But, for a moment, let us examine the possibilities of having a communication with the robot by radio signals.

If you are familiar with the antiaircraft missile systems used by the Department of Defense, then no doubt you will be concerned if we do not also point out that in these fine systems, the information concerning the target can be obtained by the ground station, independently of the missile. The missile is really a controlled robot projectile of the robot ground station complex.

The point is that two way communication with a robot is essential. We send commands, we get feedback information from the robot's sensors. The computing section of the system analyzes the data, modifies the commands in accord with that data and sees to it that everything works as it should, for as long as it should, in the proper sequence, and with the proper timing. The computer coordinates various machines and perhaps other robots in doing a job so, the computer must get feedback from each robot.

Think of the amount of information a complex operation must have moving over a communications system! This information must not be distorted or modified or lost in the process, and it must be very visible to the computer in whatever noise levels may be present in that communications system.

We are constantly amazed at the human capability to distinguish desired information in an environment of unde-sired information. For example, three persons may talk at the same time and you want to follow the intelligence of just one discussion. The human mind, once given the command to listen to Joe, turns its attention to what Joe is saying, exclud-ing the other voices as far as interpreting intelligence from them. The head turns to give Joe's voice the best reception and to minimize the reception of the other voices. The mind then instantly develops a pattern of information based on snatches of words which it recognizes clearly. The mind will actually fill in words and phrases and sentences which are incomplete due to competing noises or distortions. If the mind gets lost, it commands the tongue to activate and ask "What? What did you say?"

So it is with robotic communication systems. By using the binary code and making words out of the infinite number of patterns of these symbols, sounds, or signals (in which-ever form they may appear), one might loose some elements and still not lose the message. That is a required and desired level of operation for robot machines.

Suppose that a signal comes which has the following form 110110110100. The zeros underlined for emphasis mean that no pulse of energy appeared at that time. The question then is whether a pulse of energy was there and was lost, or whether a zero was actually intended at that time. If we suppose that the computer receiving the signal can look at the word received and calculate that for this word to fit into the intelligence pattern of the phase, the first zero has got to be a one. It makes a one out of that space and the word has meaning in its phraseology. Everything continues in a fashion.

With humans, the loss of letters—say with a Morse code message—can easily be filled to make sense. Suppose you receive the following; Th__ __an was tall. You supply the missing letters to get "The man was tall."

MULTIPLEXING IN COMMUNICATIONS SYSTEMS

Multiplexing means conveying many different items of information on the same path. If we use a hard wire system we know that we can send many different types of signals over a single pair of wires, and each signal, or set of signals, can be representative of information from a single source. Thus we can have, with multiple signals, multiple information from many sources sent over the same pair of wires.

Some elementary means of conveying different items of information are to use a modulation procedure, or use a time division of the use of the lines. In the first case, since, usually, frequencies are involved, we think of this system as frequency multiplexing. The second system, as you would suspect, is called time multiplexing.

Examining frequency multiplexing further we find that we can send many items of information simultaneously, since we can modulate a carrier (be it wire, radio signal, light beam, or whatever) with many different signals simultaneously. An example is our regular broadcast radio system. We know that the musical tones of different instruments—say, for example, the piano, saxophone, clarinet, drums, and a vocal—can all be sent at the same time and we humans can distinguish each instrument by its tone. For control purposes it was a practice some years past—and still is in some systems—to send tones denoting commands or information and separate them electronically at the receiving end with audio filters low frequency rf filters. It is understood that a command can be an off-on or present—absent type of thing. The presence of a tone can cause a switch to close and some electrical or electronic function to be performed when this happens. The absence of the tone will mean switch open and a cessation of that functional operation.

It is also understood that to convey a changing bit of information, such as we might need in a robotics application, that we cannot have just an off-on type of signal. The feedback from sensors must be such that they can be varied in some way to produce information about slight changes in position, or values of the object being sensed.

The tone transmission system can vary the pitch of the tone slightly around some nominal value, and a slight change can convey the information we desire to be sent to the computer decoding section. The decoder must then change a variable pitch, or tone, into some kind of machine signal such as off-on pulses which the computer can then use to do its assessment and evaluation and computation with.

So, we are now looking at bands in the tone spectrum over which signals might be varied to convey changing information. These bands all might be sent simultaneously and so we then ask ourselves if there is a limit as to how much information we might send simultaneously. We can reduce the band widths by going to higher and higher tone or signal frequencies. We find that there is a limit to how many tones we can transmit simultaneously and still recover the desired information. Scientists who have studied this problem estimate that eight audio channels is about the limit each channel getting only 12% of the available 100% modulation capability of the transmitting system.

Think of what this means. If we are dealing with a single robot, which has many self contained feedback functions, and some kind of internal computing and directing system, eight feedback channels to some outside and coordinating might be sufficient. But if that computer is directing many robots, and needs a lot of information from each to properly effect that coordination, then eight channels is certainly limiting! So we look at Time Division Multiplexing.

It seems right away that this has some advantages. One particular advantage is that each channel can get the full 100% modulation since it is singly transmitted. When the information from one channel has been sent, then the second channel's information is sent, and so on. If we can send the information fast enough, we can send lots of channels, each getting the full operational capability of the transmitting system, and, as far as the practical operation of the robotics system is concerned, if the information is sent fast enough, it won't even know that it is getting its data in spurts.

In a way, a computer already operates with this concept of time division multiplexing, for it has common busses which are time allocated to various signals by a timing clock. Thus we begin to look at robots as just another I/O peripheral. If we want to go one step further, we can imagine a combination system, using time division multiplexing and frequency division multiplexing over a common carrier which should enable us to get all the information needed from one point to another to make remote operation of robots very practical.

SOME EXAMPLES OF PULSE CODE MODULATION

To summarize the previous discussion, we are interested in the feasibility of operating our robots remotely, without the use of hard wire cables to them, or operating the robots using a single pair of wires instead of multicables such as we have learned are used to operate most robot arms. Examine Fig. 7-12 wherein we see a graphic representation of some forms of pulse modulation which might be used to convey intelligence from one point to another. The carrier might be a wire, a radio transmitter, a light beam of any frequency, or even sound, which may extend from the audible range into the supersonic. We see how a signal might be modulated by extending or shortening the time of existence of the pulses. This is called pulse width modulation. Notice that some of the pulses graphed are wide and some are narrow and realize that a single pulse will be varied about its nominal width with the signal we desire to convey. If, for example, we wanted to transmit just ones and zeros, we might set up a system where a wide pulse represents a one and a narrow pulse (narrower than the nominal value) represents a zero. Transmission, then, of computer readible intelligence would be just long strings of wide pulses and narrow pulses to make up computer words. The diagram does not reflect accurately, the number of channels being transmitted. If, for example, we use a five bit word for commands and feedback, we would have only one channel of communication. We have to define a channel as being the route to a single operating unit. Thus in the case being discussed we would have only one channel, which might mean control of one one robot. We could imagine that we would send many sequential lists of five bit code groups and thus, using time multiplexing, control many robots in this manner.

If we restrict ourselves to the width-per-pulse as a channel, as illustrated, then we find that five pulses represent five channels as shown. The width variation might be interpreted as ones and zeros as we have stated, or by varying the width in a sinusoidal manner we could recover tones and tone variations which could represent variables in the system. The manner of varying the width of a pulse, such as illustrated, is left entirely to your own imagination and requirements, one might recover saw-tooth or square-wave or exponential waveforms from the varying tonal modulation. This tone modulation or variation might be a real variable on a rf carrier, or it might just be a recoverable value from the varying rf frequency.

There is a difficulty. If we lose a pulse we have lost a channel. If the system, somehow, is able to cause pulse existence shortening or lengthening, we have spurious information being generated. If rf is used as the carrier, it may take some bandwidths to accommodate the variation in pulses. However, on the good side of the situation, this system has been used successfully in commercial applications and in hobby applications where it is used to control radio-controlled model aircraft. It is relatively easy to design a decoder to separate wide pulses and narrow pulses.

In the second possibility the pulses all have the same width and are all transmitted sequentially. It is the relative position of each pulse with respect to a starting time for the "train" which governs its information carrying content. As shown, we assume that when no information is being transmitted, each pulse has a definite existence time, following

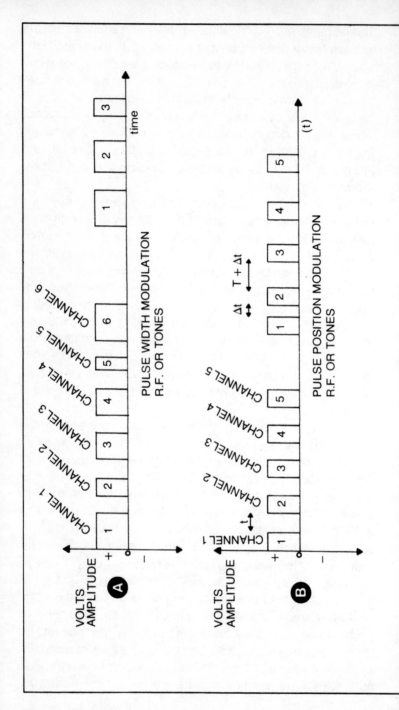

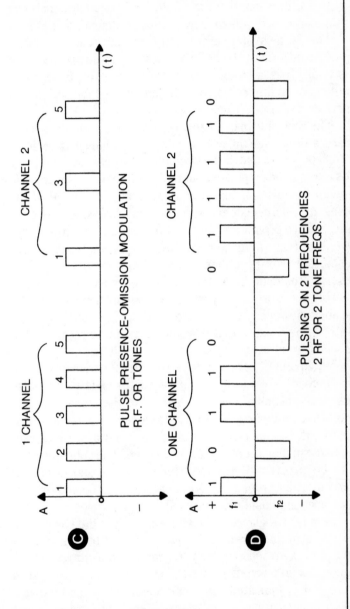

Fig. 7-12. Forms of pulse modulation. Arrangements of pulses to convey information.

some type of input starting clock signal. Since each pulse can be varied about this nominal existence time, being either early or late one can relate these positions to coded data. Early being one, late being zero. With robots such signals can also mean up arm, down arm, and neutral arm position.

This is the system of transmitting command and control signals used with the GARCAN robot discussed previously in this book. It is a simple and reliable and well understood method of sending commands. It can be used for both off-on functions and proportional functions.

The code shown in C can be a direct representation of the binary words used in computers. One simply has to have enough pulses to match the computing system used; i.e., an 8 bit system needs 8 pulses, but one might have 12 or 16 or 32 or even 64 pulses if that kind of accuracy is needed. If the pulses are microsecond or fractional microsecond existence units, and are separated in time by fractions of a microsecond, it would not take long to send a 16 or 32 pulse word or train of pulses. This word could be repeated over and over, just as a message could be repeated to insure accuracy. There is probably some probability mathematics which would indicate just about how many times, on the average, one should repeat a word to be assured that it would be properly received, but we won't go into that here.

We find in C a system made to order for transmission of binary code. We assume that signal presence—or pulse presence—means a one and absence of the signal means a zero. Therein lies our difficulty. If we lose a pulse in transmission, the computer on the receiving end might interpret this to mean that a zero should be written. The computer thus might be incorrect! But, as we have stated, redundancy might overcome this problem to some extent, and computer sentence analysis could also be used.

Finally, we glance at D wherein we find the basic radio teletype system of frequency diversity used to convey ones and zeros. Two tone frequencies can be transmitted over a common radio frequency carrier, or other type carrier, such as previously indicated. One tone means a one, and the second tone means a zero, as shown. Since the system must always have one or the other tone being received, any absence of tone means an error which the computer can recognize. As we know, this system has been found to be highly accurate and reliable in the transmission of messages. It would be relatively easy to apply this system to robotics applications and, using filters, and combination or pairs of tones, one might control many robots, or robotic functions over a single wire pair, or rf frequency.

SOME DESIGN CRITERIA OF TELEMETRY SYSTEMS

Measuring the responses of a robotics system by remote means is similar to getting information from an experimental aircraft or rocket in flight using radio systems and sensors at various critical points in the system. Thus, a few moments consideration of telemetry design criteria can be helpful in our study of establishing communications with robots.

We know that radio telemetry systems have been used in aircraft to measure such things as position, velocity, acceleration, and aircraft attitude with respect to some gyroscopic coordinates. Also, sensors are used to determine the amount of vibration or flutter at various points on the body and the pressure at various points internally and externally. They also measure temperatures, strain, stress, fuel flow, cosmic ray counts, ionization, static build-up, and voltages throughout the aircraft. Of course such measurements can be made on other vehicles as well. Our space explorers are robots that send back pictures, and after testing and sampling distant planets, information on soils, atmosphere, and such is also sent.

When we consider the return of data, we must also consider the accuracy of that data and whether it meets the design and operating criteria or not. Data from such systems may vary quickly and considerably, or it may vary slowly and within small excursion limits. If we are looking just for general information then, perhaps, accuracies within, say, 15 to 40 percent may be used and acceptable. But if we really want tight monitoring then we will want accuracies within, say, ½ to 2 percent and in some cases, as in a robotics system where positioning is being monitored, we may want accuracies from

0.1 to 0.001%! Of course we would like absolute accuracy; zero percent error if possible. But due to system limitations we just can't reach that optimum.

When we are interested in some sensor value which may vary at a sinusoidal rate, it has been found necessary to sample that quantity with at least 3 samples per cycle, and it has been said that you must have at least six samples per cycle to reconstruct a varying sine wave with any degree of accuracy. Of course, if you want to transmit data that is constantly being measured, as you would if you used a frequency varied by the sensor, then you should have total information instantly, concerning that measured unit. If you use a pulse sampling system wherein the pulse voltage amplitude is then converted into digital (binary) words which can be used by a computer to reconstruct the wave being sampled, there is always the possibility that the sampling system might miss something—some variation—if that sampling is not conducted frequently enough.

In accepted telemetry systems which use tone or low rf frequencies to modulate a carrier (usually an FM carrier to reduce noise and distortion), the system is called an FM—FM system. This means frequency modulation of the oscillator which, in turn, is frequency modulating a carrier. The oscillator, of course, is being modulated or varied by the sensor measurement unit. If one wishes to think in terms of computerized handling of data, then one thinks in terms of receiving pulses which form groups of binary computer words as we have already mentioned. Analog to digital circuits are available and can be used in such an application. The accuracy will be governed by the number of samples-per-second taken of the measured or sensed quantity. A representative telemetering system of the FM—FM type is illustrated in Fig. 7-13.

In the illustration we see some sensors which cause the frequency of the oscillators to change according to the movement, strain, velocity, or whatever is being sensed. This signal is then used to FM modulate the transmitter. It is sent to the receiver, there separated by band-pass filter sections converted into varying levels of plus or minus dc voltages

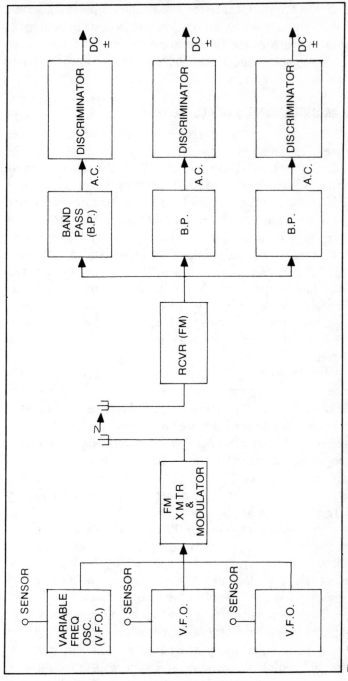

Fig. 7-13. A representative FM-FM telemetry system.

273

which are proportional to the instantaneous position, strain, or velocity of the sensed unit, and then sent to analog-to-digital converters and thence to computers. Of course these signals might be used in their analog form to control a remote robot or whatever.

COMMUNICATIONS SIGNALS NEEDED

It is essential that regardless of the type of carrier which conveys information to the robot, this carrier must also be capable of returning signals from the robot to its control station. We think of a simple, proven example. If one has a guided missile whose control station and computing station are on the ground, and the mission of this missile is to destroy enemy aircraft in flight, then the missile should, under some circumstances, be able to inform the ground station how close it is to the aircraft it seeks as prey. When it is close enough, the ground computer will determine an exact time to detonate a warhead for destruction of the enemy aircraft. The signals might also be used to drive pen recorders, printers, or illuminate video displays, as desired, in the operation being conducted.

It is interesting that there are some 18 frequency bands, made up of frequencies from roughly 350 Hertz to some 70,000 Hertz which can be used in FM telemetry. One must make sure that he does not use two bands which are harmonically related, as that plays havoc with the system's ability to give out good data. Variations in frequency are relatively small, usually about 7.5%, and of course, the higher the band frequency, the higher the frequency response capability of that particular sensing section can be.

Thus, when we consider the use of robots which are not fixed in a given location, we think in terms of systems of control which can radiate a roomful of signals which can be received at any place in the area or room, and getting back responses from the robot in the same manner.

The reference system in which all movements are determined is the instantaneous position which they might have at that moment, probably with respect to the four walls and floor and ceiling of the room, and with respect to whatever objects may be within the room, even to a consideration of other people who also may be moving! The feedback sensor system must give you (or your computer) this data. In an automated robotic system with free wheeling robots in action, some computer somewhere must know everything about everything and keep this knowledge accurate and current!

CURRENT SOCIOLOGICAL THINKING ABOUT ROBOTS

To many people, not well enough informed about automation and robotic technology, or incapable of completely understanding just how these machines work, the existence of robots present them with a threat. To some it is a job-loss threat, to some it is a much more serious threat such as; "Will robots take over and rule the world making us slaves?" We may grin about this, but, be assured, such thinking exists in no small dimension. The Robotics International Organization defines a robot as "A reprogrammable, multifunction manipulator designed to move material, parts, tools, or specialized devices through variable programmed motions for the performance of a variety of tasks." One can see, in this definition, a description of a machine and that may be a big help to industry when they plan the utilization of these devices. It may help out with the psychological problems of people.

There are many fantasies concerning robots and these become more prevalent when we consider the robot in some kind of human form. For some reason, that seems to be what the average person would like for his robot to have. One of the leading Science Fiction writers, Isaac Asimov has written of a situation where a mother becomes jealous of a robot who was assigned a baby-sitting task over her daughter, in the story *ROBBIE*. In another story he described the relationship between a wife and a robotic butler. Movies have been made which show humans falling in love with robots. Why people would be attracted to robots that are, to all appearances, duplicates of themselves, or of "beautiful people" can be the subject of another study. We just pose the idea here for consideration.

In all pseudo-science stories concerning robots we find that sooner or later, the "Laws of Robotics" are derived or stated. These laws are supposed to be the statements which govern the limits of robotic action. Isaac Asimov formulated and first wrote the Laws. One of these is paraphrased "The robot shall not harm his maker." Designers were supposed to insure, in the circuitry associated with the robot, that this Law was embedded and used as a control on the robot's operations. Now we find that there is an effort to state some Laws for industrial robots, and since these, in a way, reflect thinking concerning the use of industrial robots, we should examine them. As stated by Neale W. Clapp of Block Petrella Associates, one set of such Laws are:

(a). Organizations may not install robots to the economic, social, or physical detriment of workers or management.

(b). Organizations may not install robots through devious or closed strategies which reflect distrust or disregard for the work force, for surely they will fulfill their own prophecy.

(c). Organizations may only install robots on those tasks which, while currently performed by men, are tasks where the man is like a robot, not the robot like a man!

This decade is the beginning of the age of robotics, and thus, while we are learning how to make them operate, and what they can do, and how to make them better from a design and operational standpoint, perhaps the sociological aspects of robotics showed also be a prime area of consideration. This could also influence the design, installation, development, and operation of these types of machines.

HOW SOME EXECUTIVES DETERMINE THE NEED FOR ROBOTS

W.R. Tanner of Tanner Associates has stated some rules of thumb by which an executive might determine if he is on a correct course when he installs robots in his installation.

(a). Avoid extremes of complexity

(b). Operations must be orderly and systematic

(c). Remember, robots are generally no faster than people

(d). For short runs use people, for very long runs use fixed automation (e). If it doesn't make dollars it doesn't make sense

(f). One is not better than none

(g). If people don't want it, they won't make it.

Refer to the Spring, 1980, issue of *Robotics Today* magazine for Mr. Tanner's discussion.

Some of the ideas expressed in the rules seem suitable, and others are open to question. We would question f as it seems that a detailed consideration of the application might be in order before making a judgment such as stated here. If a robot can be operated 24 hours a day (not considering down time), and if it requires no breaks or rest periods, and if it can be adjusted to perform new tasks with relative ease, then perhaps just one robot in an installation can add up. You will determine, from your own experience and knowledge, which of the other rules might be appropriate, but we would venture to suggest that some debate might be worthwhile on some of these rules. We have included some of the previous rules and definitions to illustrate that industry, as well as users of other types of robotic equipment, seem somewhat frustrated when they really begin to consider such machines in the hierarchy of human-machine relationships. When we think of an automobile, we simply think of transportation. It can be with or without comfort and conveniences, and it can be cost effective or not. It can be status making or demonstrating. But we know where and how to think of that kind of machine, both in a pleasure sense, and in a profit producing capacity. We also know how to relate the human element to the machine, even if it is a variation of the automobile, such as the truck, tractor, tank, or whatever.

Perhaps we must come to the same type of thinking with robots? Is it possible that one status symbol of the future will be the number of domesticated robots one has around the house? It is possible that the industrial profit margin will be controlled by the number, type, and efficiency of robot workers. It is possible that one might find a completely new type world in which the human must be re-programmed to be compatible with the robotic machines which he previously programmed?

As I write this book I wonder what the future of writers will be. Will a robot typewriter investigate the many areas of research by simply sending signals over the telephone lines to libraries and other repositories of knowledge; analyzing and assembling and developing material and then spewing out pages of text? Perhaps it would write novels, based on tried and true plots, or screen scripts? Man may tend to forget, and a computer can memorize almost forever. Man may be emotional in rendering judgments, a machine will just consider facts. Man may be unable to render perfect decisions, due to his inability to retain or consider all facts needed to get the absolute picture of background, trends, or whatever. A computer, with the new technology making possible a larger and larger memory capability, and with the computer ability to solve equations relating to the decision making process that baffled man in years past, we, even currently, note how many humans are turning to these machines for the answers to their problem.

In any event, we need to know as much as we can about these robotic computer-controlled things and to this end we now continue into the next chapter where we consider first another industrial robot, and then some aspects of giving robots eyes. You might even begin to think about the situation we have now, where most robots are blind. What will the ability to see mean to these machines?

The ASEA
Robot System

One unique feature of this robot is that it is an all electric system. As illustrated in Fig. 8-1, you can see the large, specially designed electric motors on each side of the arm at the shoulder position, and one to the rear of the arm section. Also visible in the illustration is the control panel which governs its operation. At the right end of the arm itself is an attachment disc which permits use of a multitude of tools or devices to do the various tasks for which it is designed. Notice that while the arm itself is not directly extendable, reach can be obtained by use of the electric motors at the shoulder position, or by causing the arm to bend at the elbow rotary joint. This is very similar to the human arm.

When we examine such an arm and realize its strength and capability, it brings to mind a vision of a sleeping giant and helpers as illustrated in Fig. 8-2. In this illustration we attempt to show that only a portion of a robotic capability, perhaps, is being used to assemble or adjust some objects brought to it by, in this illustration, elves, who position and hold the workpieces while the arm performs its operation on them. Then they move on to the next robotic station. This, of course could be some type conveyor system, but one which has a positioning capability at the end where it meshes with the arm operation.

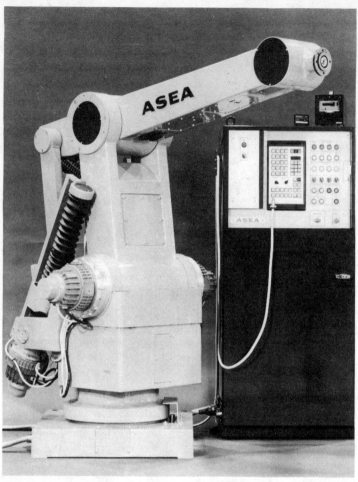

Fig. 8-1. The ASEA robot for industrial applications (courtesy ASEA).

As to the giant itself, note that only a small portion of the possible computer brain has been used, and that a second arm remains motionless. In some applications there is a need for a second arm, or even a second pair of arms and some robots have been designed to have these appendages. Underwater repair of cables or oil lines use such robots to perform working tasks. Our giant gets his orders by telephone or more accurately, by electronic-electrical signals. Maybe it's a little fanciful but it is intended to illustrate a point. The ASEA robot

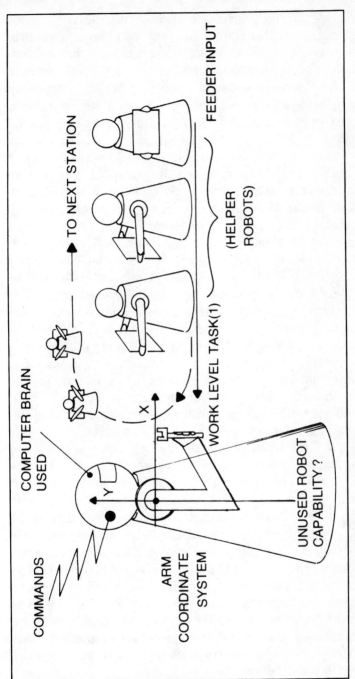

COMMANDS

COMPUTER BRAIN USED

ARM COORDINATE SYSTEM

UNUSED ROBOT CAPABILITY?

Y

X

TO NEXT STATION

FEEDER INPUT

(HELPER ROBOTS)

WORK LEVEL TASK(1)

Fig. 8-2. One arm benign giant (robot) and helpers.

is made by the Electronics Division of ASEA, Västeräs, Sweden. More than 600 of these type robots are being used in various types of industrial applications in more than 20 countries. The tasks which it performs range from moving a stainless steel sink around a polishing brush, to delicately applying liquid adhesive glue to a car top, to applying a multi torch welding unit to car bodies. This robotic arm has six degrees of freedom to do the tasks assigned. Notice in Fig. 8-1 how well sealed the whole unit is, and that makes it possible for it to operate for long periods of time in dirty environments where grime and other contaminants may exist. In one de-burring application a vicious stream of tiny metallic particles is generated during the working process. The arm must not be affected by them. Each electric drive unit includes a complete servosystem with a dc motor, resolver, for positioning indication, and a tachometer for controlling its speed. It is said to have a quiet operation.

What we gain by examining this unit is a further understanding of just what an industrial robot consists of, how it is programmed, how it operates, and what it can do. Since this arm is electrically driven, it also can provide a giant example for the hobbyist or experimenter as to how electric motors are, and can be, used in such an applications, and how they are controlled and regulated. Smaller motors, of course, would be used in smaller arms.

In the description which follows we must realize that the robot arm is described as having a lower arm section which is that section connected to the base pedestal, and an upper arm section which is the horizontal member attached to the lower arm section. The robot can rotate about its base pedestal and the motions of the arm segments will be angular about their pivot points, but this angular motion can result in an in-out motion of the end wrist to which various tools and devices can be attached. The wrist is the small circular section at the right end of the horizontal upper arm in Fig. 8-1. Its rotation is perpendicular to the rotational up and down movement of the upper arm. In this robot the designers have used an ingenious system of linkages to accomplish wrist rotation. Now let us learn some details of this robotic arm.

DESCRIPTION

The ASEA industrial robot system is made up of three main parts:

Control system
Measuring and servosystem
Mechanical system

The control system consists of the computer, memories, inputs and outputs to control the robot and interlocks from peripheral equipment, and functions to control the servo-system of the robot.

The measuring and servosystems include servo amplifiers and dc motors with tachogenerator feedback. Position regulation is by means of a cyclic resolver system consisting of a resolver with associated supply and decoding circuits, and a position regulator.

The mechanical system includes the robot and the transmission which converts the rotation of the motors into the required motion. The following motions are available:

Rotation	(ϕ)	entire robot rotates about its pedestal
Out/in motion	(θ)	lower arm moves
Up/down motion	(α)	upper arm moves
Wrist bend	(τ)	the wrist bends up or down
Wrist turn	(v)	the "cuff" of the wrist rotates

Various additional units (options) can be connected to the control system of the robot:

A programming unit, which is used to program robot motions

A tape recorder unit, used to store programs on magnetic tape cassettes

Standby battery supply, ensures that the contents of the program memory are maintained up to 45 minutes after loss of supply

Memory module to increase the size of the program memory

Test panel for servicing and troubleshooting

Solenoid valves for gripper operation
Gripper

The control system is based on a computer, as shown in Fig. 8-3, which illustrates the main communication paths in the system.

A control program is stored in the memory. This program, which is fixed, tells the computer how the various robot instructions and control functions are to be performed. The operator and control system communicate via the control panel and programming unit. The computer regularly scans the state of the controls and reads out acknowledgement

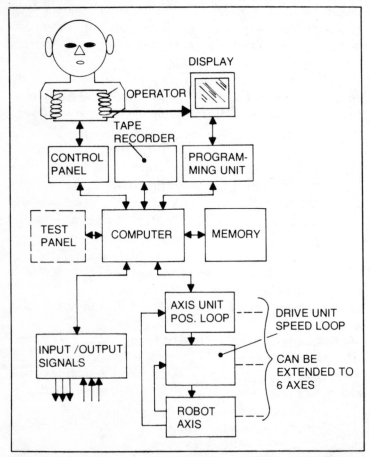

Fig. 8-3. The block diagram of the robot system of ASEA (courtesy ASEA).

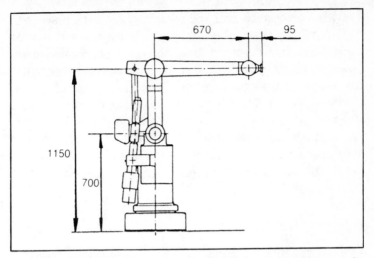

Fig. 8-4. Dimensions of the smaller industrial robot arm with a 6 kg capacity (courtesy ASEA).

orders by switching lamps on or off. The computer monitors and controls input/output signals between the process and the control system, and reads out orders to the axis units if travel is required on any axis.

The operator programs the robot with a particular pattern of motions by means of the programming unit. Each move is commanded by the operator using jog buttons and the position and speed are stored in the computer memory. The conditions for process signals (e.g., for gripping tools and external acknowedgements or interlocks) can also be programmed and stored in the memory. When the program is complete the operator can operate a control on the control panel to order the computer to store the program on magnetic tape. The program can then subsequently be read in from the magnetic tape and stored in the memory of the computer. A robot program stored in the memory can then be started and continuously repeated or stopped from the control panel on the order of the operator.

We have mentioned that a robotic arm should be capable of doing work, and thus it requires power. ASEA makes two robotic arms seen in Figs. 8-4 and 8-5. One is much smaller than the other (IRb-6vs IRb-60). It is now of interest to use to see what some technical specifications are with respect to weight handling capability, degrees of freedom, and gripper capabilities. In many applications the speed of arm movement can be very important. Thus we also examine this capability. Remember that as the size, and thus the inertia, of an arm increases, more power is required to make it move quickly, and also much more attention must be given in the servo-system to the damping function to insure that the arm will not overshoot its required positions, and that it will not hunt or oscillate about an end-point position.

Power

Robot	IRb-60	IRb-6
Permitted handling weight including gripper:	6 kg	60 kg
Maximum gripper length at above handling weight:	200 mm	400 mm

In order to calculate the permissible handling weight when the gripper length exceeds 200 mm or 400 mm as the case may be, the following mechanical data apply:

Maximum moment of inertia	2.5 Nm²	100 Nm²
Maximum static load	12 Nm	240 Nm

The robot is available with the following degrees of freedom:

Arm motion: rotation	340°	330°
Arm motion: radial	± 40°	+50° to −20°
Arm motion: vertical	+25° to −40°	+10° to −55°
Wrist motion: bend	± 90°	+75 to −120°
Wrist motion: turn	± 180°	± 180°

Gripper functions:

Two independent solenoid valves housed in the upper arm, which can be operated from the keyboard of the programming unit, are available as an option. There is also a four-pole electrical outlet in the upper arm; among the intended applications of this is the supply of more advanced grippers with search functions.

Weight of robot:	125 kg	750 kg
Repeat accuracy at wrist:	±0.20 mm	±0.40 mm
Speeds:		
Arm motion: rotation	95°/s	90°/s
Arm motion: radial	0.75 m/s	1.0 m/s
Arm motion: vertical	1.1 m/s	1.35 m/s
Wrist motion: bend	115°/s	90°/s
Wrist motion: turn	195°/s	150°/s

Every robot arm has a working envelope or area through and in which the arm can be programmed to move. We show the profile of the IRb-60 in Fig. 8-6. Again notice that the dimensions are stated in metric units.

Description of Motions

The basic version of the robot has three degrees of freedom for positioning the hand in space. The three main

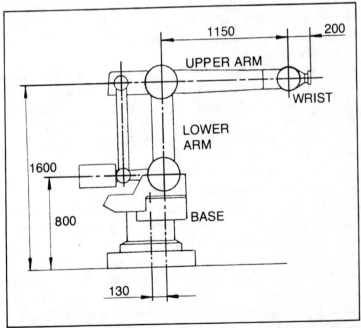

Fig. 8-5. Dimensions of the larger industrial robot arm with a 60 kg capacity (courtesy ASEA).

coordinates are a rotary motion (ϕ) and two arm motions in the vertical plane, performed by a lower arm (θ) and an upper arm (α). All three main motions are position-controlled by means of dc motors. A further two degrees of freedom are available as options, a wrist turn and a wrist tilt version.

These motions are available either with dc drive or with pneumatic rotary cylinders with mechanical limit position stops. In the basic version these motions are mechanically blocked.

The drive unit for the rotary motion is rigidly mounted in the pedestal and drives the body around the pedestal column through a reduction gearbox with a very low ratio. See Fig. 8-7.

The drive unit for all coordinates except the rotary motion (ϕ) are mounted on the rotating body. The arm motions are provided by ballscrews which actuate the arms via links and levers. The lower arm (θ) moves the wrist horizontally in and out. The motion remains completely horizontal since the α-motion compensates for the θ-motion by means of the control system. In the same way, the motion of the upper arm (α) is compensated so that its motion moves the wrist vertically up and down.

The wrist motions are driven via a system of linkages which is designed so that the wrist coordinates will assume the set angle to the horizontal plane regardless of whether the θ- or α-coordinate changes. The drive units for the wrist motions are mounted on the frame at the lower bearing point of the lower arm.

The wrist is also designed in such a way that tilting always takes place in the vertical plane.

The driving system for this arm consists of dc electric motors. It is easier to design a driving system to produce variable speeds and torques using dc motors than it is using ac motors. When using ac one must have modulators, and demodulators and phase comparers and such. With dc motors one simply uses high powered transistors or similar solid state drivers. Since signals from computers are dc this also makes the whole system more compatible, electrically. In Fig. 8-7 we show the drive unit locations and some defini-

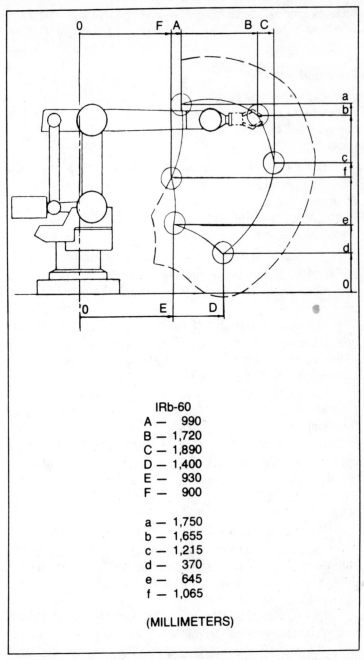

IRb-60
A —   990
B — 1,720
C — 1,890
D — 1,400
E —   930
F —   900

a — 1,750
b — 1,655
c — 1,215
d —   370
e —   645
f — 1,065

(MILLIMETERS)

Fig. 8-6. The working profile of the IRb-60 robotic arm (courtesy ASEA).

tions. A unit as stated here is an electric motor. Inspect the wrist motions and notice that both turn and tilt are accomplished.

Drive Unit Analysis

The drive units consist of:
Motor unit
Mechanical transmission
Servoamplifier

The motor unit of the basic version is made up of the following components:
Dc motor
Resolver for position indication
Tachogenerator for speed regulation

The motor unit for the wrist turn or tilt motions can be replaced by a pneumatic turning cylinder, two versions of which are available, one for 90° turn and the other for 180° turn.

The transmission is of two types, the harmonic-drive gearbox or the ballscrew unit. Each motion is described in the following pages, which also deal with the transmission, that is the way in which the motions are achieved.

The servoamplifier is installed in the control cabinet together with the rest of the electrical equipment.
Pedestal:

The pedestal is in the form of a baseplate with a cylindrical top; it is an aluminum casting. The cylindrical part houses the motor unit and a harmonic drive gearbox for the rotary motion (ϕ). See Fig. 8-8.
Body:

The body is coupled to the output shaft of the ϕ-motion drive unit. It is also mounted on a bearing in the pedestal and rotates on this (the ϕ-motion). Mounted on the body are the drive units for lower arm motion (θ-motion) and upper arm motion (α-motion).

The electrical wiring from the control cabinet is connected in the pedestal by means of connectors and passes from the pedestal to the body through coiled flexible leads between the pedestal column and the body.

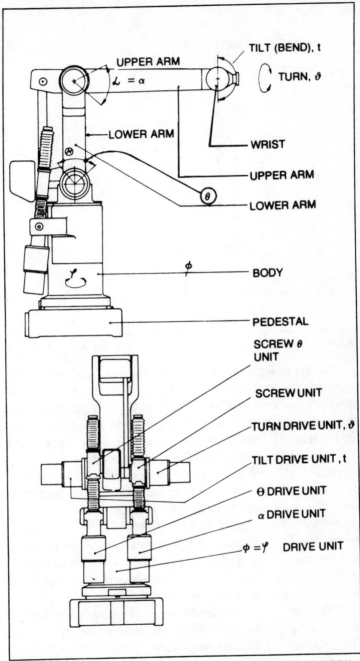

Fig. 8-7. The drive units of the IRb-60 and some definitions (courtesy ASEA).

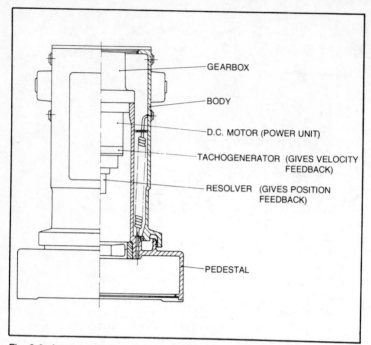

GEARBOX

BODY

D.C. MOTOR (POWER UNIT)

TACHOGENERATOR (GIVES VELOCITY FEEDBACK)

RESOLVER (GIVES POSITION FEEDBACK)

PEDESTAL

Fig. 8-8. A cut-away view of the robot's pedestal (courtesy ASEA).

In this illustration we see that the left hand side is the enclosed part which shows how the body and pedestal appear as you view them, such as in Fig. 8-1. On the right half of the illustration the internal location of parts and devices is shown. Notice that the tachogenerator which we call a tachometer which is really a small generator which produces a voltage output exactly proportional to its speed and which is used in the damping of the system, is an integral part of the motor unit. The resolver (we sometimes call them selsyns or synchros) will give a signal (ac) the phase of which is directly proportional to the instantaneous angular position of the body with respect to the pedestal. This is used for position feedback in a servosystem. A dc resolver is a potentiometer and it is sometimes used in servosystems.

Next we begin an examination of the moving sections of the robotic arm and the gripper unit. Notice that small lever arm sections are located near some rotating joints so that a motor driven screw can exert the force necessary to move the arm section. The manner in which this driving force is exerted is easier than using the motors directly connected to the joint through a complex gear train as necessary in a direct-driving system. (In many cases direct-drive *is* used, and very successfully.) The screw-type lever-action motion shown in Fig. 8-9, approximates a hydraulic type piston motion. But the reserve oil tank, the valves, and the pressure pump and use of oil are eliminated which may be a cost factor. Speed of motion is reduced over that of a hydraulic system, and perhaps some power capability is lost when using electric motors in this arrangement. Look at Fig. 8-9.

Mounted on the body is a bearing bracket to which the lower arm is fixed and in which it pivots. The motion of the lower arm (θ-motion) is obtained by means of a drive unit consisting of a motor unit and ballscrew transmission; these are rigidly mounted on the body. The motion of the ballscrew is transmitted to the lower arm via a lever pivoting at the ballscrew and attached to the lower arm. The drive units for

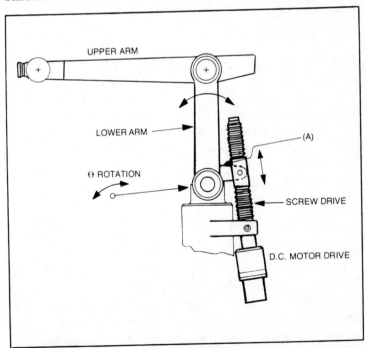

Fig. 8-9. The lower arm motion system of the ASEA Robot arm (courtesy ASEA).

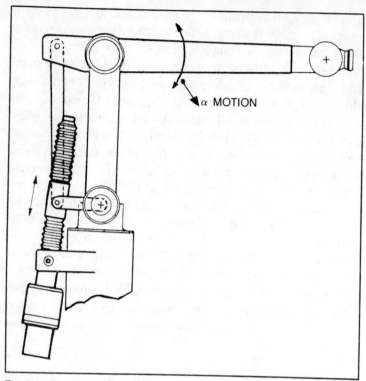

Fig. 8-10. The upper arm motion system and drive assembly for ASEA robotic arm (courtesy ASEA).

the turn and tilt motion of the wrist are mounted on the lower arm around its lower turning centre, see Fig. 8-11.

The upper arm description and operation are as follows: The upper arm is fixed to, and pivots on, the upper end of the lower arm. Movement of the upper arm is achieved as follows.

The drive unit, consisting of the motor unit and the ballscrew transmission, transmits a motion to two link rods articulated on a shaft extending from the ballscrew nut assembly. One of the link rods is attached to a shaft fixed to the upper arm and the other end is attached at the center of rotation of the lower arm. Together with the upper and lower arms, these two link rods form a parallelogram. As Fig. 8-11 shows, movement of the nut assembly and one corner of the parallelogram will move the wrist up and down.

Electrical and compressed-air lines to control the grippers run through the upper arm. At the rear of the arm there is a connection for the supply of compressed air. Outlets for the gripper, two individually controllable pneumatic outputs and an electrical contact with four wires are mounted on the underside at the front of the arm.

The wrist is rigidly mounted on the upper arm. Two motions are available at the wrist, a turning motion and a tilting motion. These motions are achieved as follows. The drive unit for each motion can either be a dc motor equipped with a resolver and a transmission consisting of a gearbox, or a pneumatic rotary cylinder for turning at right angles or 180°. With pneumatic power, only one motion can be driven and the other must always be locked. The motion of the drive unit is transmitted to the wrist via a linkage system.

The linkage system operates as follows, Fig. 8-11. A linkage disc is coupled to the drive unit, which is mounted on the lower joint of the lower arm. Mounted on this linkage disc are two link arms displaced $\pi/2$ rad (90°) on the linkage disc. The other end of the link arms is mounted on a linkage disc pivoting in the upper arm. From here the motion is transmit-

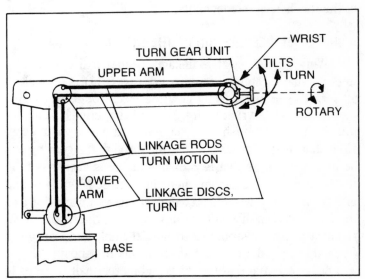

Fig. 8-11. A schematic diagram of the wrist linkage system for ASEA Robot (courtesy ASEA).

ted to a linkage disc in the wrist via two more link rods. The turn and tilt motions have separate linkage systems on either side of the arm system. Wrist tilt is achieved because the linkage disc at the wrist is fixed to the moving part of the wrist. Wrist turn is achieved by means of an angled gear unit which transfers the motion from the linkage disc to a turning disc, free to rotate in the wrist. The angled gear unit is adjustable so that subsequent adjustments can be made in cases requiring very little backlash.

Figure 8-11 shows the linkage system in the lower arm and upper arm. In the basic version the disc for the tilt motion at the lower joint of the lower arm is mechanically locked, so that the wrist keeps a given angle relative to the horizontal plane regardless of arm motions. The turn motion is mechanically locked at the wrist itself.

Mechanical stops are rubber composition. There are mechanical limit position dampers for all motions, to prevent damage to the mechanical components of the robot if, because of faults in the control system, the robot motions should exceed the working range. The dampers are of rubber and are designed to damp the full motion speed without deformation of the mechanical components.

A SIMPLIFIED DESCRIPTION OF THE LINKAGE MOTION

While ASEA uses two linkages from one disc to another in the actual machine, the drawing (Fig. 8-11) is difficult to analyze to see just exactly how the discs rotate to cause a movement of the wrist. In Fig. 8-12 is a simplified look at how this can be accomplished. If you follow the arrows and consider that the motor output rotation through the gearing will be limited to somewhat less than 90 degrees each way from the neutral position shown, then you can see how the output disc to which the wrist is attached will move up and down as the motor driven disc is turned. Of course, the motor might just have a small lever output to move the linkage up and down (A) and not need a disc (B) at all in this first position.

It is essential that the discs and motor shaft be fixed to the robot's body so that no movement up or down, left or right will occur. If they move physically, then they cannot transmit

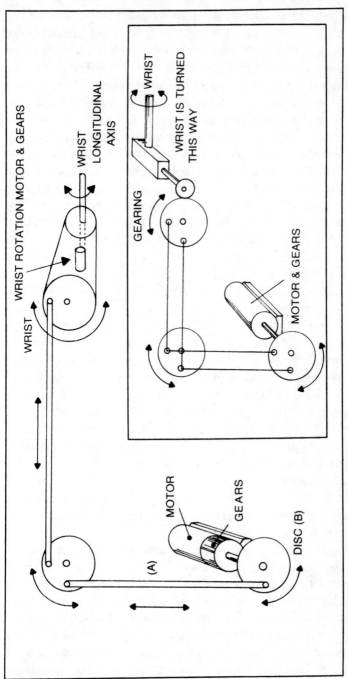

WRIST

WRIST ROTATION MOTOR & GEARS

WRIST LONGITUDINAL AXIS

WRIST

WRIST IS TURNED THIS WAY

GEARING

MOTOR & GEARS

MOTOR

GEARS

(A)

DISC (B)

Fig. 8-12. A simplified diagram of the ASEA linkage motions.

297

the rotational movement which we need at the wrist. Also, you can consider another set of such discs and another motor and some additional linkages which can be so positioned that when this second motor is rotated each way somewhat less than 90 degrees, it can cause the wrist to turn sideways, back and forth. Thus we have the movements needed for back and forth and up and down. The rotation or turning of the wrist about its own axis can be accomplished with an integrally installed motor at the wrist position, said motor to be turned and tilted when necessary, and when activated, it can, through its own gearing, provide a rotational movement 90 degrees one way and 90 degrees the other way, of the wrist, or whatever is attached to the wrist. It would be advisable, of course, to provide electrical limit switches on the wrist so that the motor cannot overdrive it past the 90 degrees positions. Examine Fig. 8-12.

We recommend that if you are deeply interested in this linkage situation, you might contact some mechanical engineering student, or friendly professor, at a nearby university and question them about Fig. 8-11. A second source of information on linkages of various types, pantographs, etc., can be found in your local library under mechanical devices or mechanical engineering mechanisms. Ask the librarian to help you locate a good reference work if you need it.

No one has said that a robotic arm is a simple device, it just looks that way! Actually there is quite a bit of mechanical, electrical, and computational equipment installed in such a unit. Examine a block diagram of the ASEA robot system in Fig. 8-13. The legends are in English and Swedish so a brief explanation of the operation will help tie all the various blocks together.

The control electronics consist of circuit boards which make up a computer with a memory. There are also interface boards for data communication with the various input and output units, i.e. tape recorder, control panel, programming unit and axis units. The signals between the function modules of the control electronics appear at the rear plane of the electronics rack and are in the form of a data bus, memory address bus, address pointers for the various input and output units and a small number of clock pulses and control signals. The digital input and output signals for the control panel, programming unit and external process go to the input/output unit via the terminal unit. An analog reference signal, the polarity and value of which indicate the direction and speed of movement of the robot axis, is generated in the axis unit. The analog reference (command) signal is amplified in the drive unit so that it can control the dc motor.

The computer included in the control electronics can be divided up into five main sections.

●Input for reading-in data from the process (feedback)

●Arithmetic unit for arithmetical and logical data processing

●Memory to store both data read in from the process (feedback plus command signals) and the robot system control program

●Control functions to check what is to be executed

●Output for the supply of data to the process

The arithmetic and control sections consist of the function modules known as the central unit (A) and the interruption and power failure detection unit (B). These units contain a clock pulse generator to generate clock pulses to the control section and other electronic sections, a microprocessor and circuits to control a general 8-bit data bus for data transmission between the microprocessor, memories and input and output units. The microprocessor contains, among other things, an arithmetical-logical unit, an instruction decoder and a program counter. The instruction decoder interprets the instruction read from the control program of the memory and establishes the conditions for the execution of the required operation. The program counter keeps track of the order in which the computer instructions in the control program are to be carried out.

The memory, which may be either of the write/read type or of the read type only, consists of various combinable function modules, depending on its size and type. The memory is organized in memory words of eight bits each. A total of 16 k (16,384) words can be addressed.

There are two electronics modules for the computer memory: the memory unit and the PROM-unit.

The memory unit contains a 4 K (4,096) word write/read memory. The write/read memory is a semiconductor memory which stores the robot programs programmed by the operator. It defines the patterns of motion of the robot, that is, data for position coordinates and speeds, and the control of digital inputs and outputs. The second variant of the memory unit, with write/read memory only, is used to increase the memory space for the storage of robot programs.

The PROM-unit, which contains a 7 K (7,168) word read memory, stores the control programs of the systems. The memory content of the PROM-unit does not change in normal service, but it can be re-programmed with a special programming unit.

Finally there are the input and output units: these are the tape recorder interfacing, axis control circuits and input/output unit. The tape recorder interface controls data transmission between magnetic tape and the memory. The "axis control circuit" function unit receives data words which state by how many increments the relevant robot axis must move. The purpose of the input/output unit is to match words on the bus system of the computer to signals for controlling and checking panel controls and external relay and contact functions. For reference voltage supply of the measuring transducers on the robot, the reference counter and resolver amplifier function unit will be required. A test panel for servicing is also available, as shown in Fig. 8-13.

OPERATION

As Fig. 8-14 shows, the basic program for controlling the robot system is divided up into a number of subprograms. The start-up routine is always activated after switching. Among other things, this clears the registers and program counter. The clock pulse routine synchronises the computer with the 10 ms sampling period used in the robot system. Whenever the clock pulse arrives, a jump to the clock pulse routine takes place, and at the same time the conditions of a group of controls are read in. While automatic running is in progress, a jump back to the automatic routine always takes place, so that new movement data are read out during the next sampling period or at the beginning of execution of the next robot instructions. In other cases a jump takes place to the mode selection routine. This routine investigates which load is required and whether any related operator actions have been taken. If so, a jump takes place to the routine for the selected mode: if not, the system jumps back to the clock pulse routine to await the next clock pulse.

The manual routine deals with all operator actions relating to the manual mode and required for robot programming, program corrections, synchronising the robot and step-by-step running during program testing. For example, when the routine is complete, the system jumps to the clock pulse routine, except in step-by-step running, when it jumps to the automatic routine in order to execute the robot instruction.

The automatic routine selects which robot instruction is, in turn, to be executed. After this, each robot instruction has its own sub-routine; in other words the automatic routine includes a total of 16 sub-routines.

Finally there are the routines "read cassette" and "write cassette", which carry out reading of robot programs to, and transfer of robot programs from, the magnetic tape cassette. After reading or writing is complete, the system jumps to the start-up routine.

Feedback and Servosystems

Each robot axis is fitted with a feedback resolver for axis position measurement. The resolver has two fixed stator windings and one rotating rotor winding. The resolver stator windings must be supplied with two reference voltages with the same peak-to-peak value and 90° out of phase. The frequency is 2 kHz. The voltage induced in each rotor winding is filtered in the appropriate "axis control circuit" function unit. The phasing of the basic frequency of the resolver response is a measure of the position of the axis. The resolver is coupled to the drive motor shaft, and one motor revolution corresponds to 200 increments.

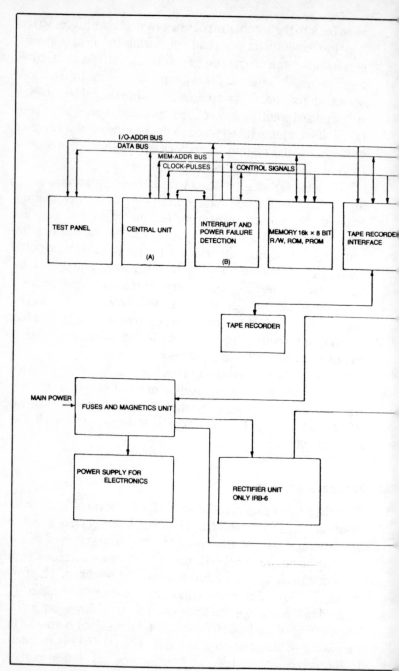

Fig. 8-13. The block diagram of the ASEA robotic arm (courtesy ASEA).

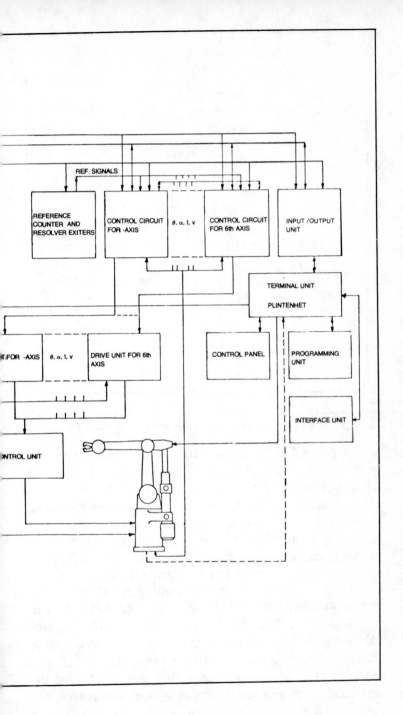

REF. SIGNALS

REFERENCE COUNTER AND RESOLVER EXITERS

CONTROL CIRCUIT FOR -AXIS

θ, α, t, v

CONTROL CIRCUIT FOR 6th AXIS

INPUT /OUTPUT UNIT

TERMINAL UNIT

PLINTENHET

IT FOR -AXIS

θ, α, t, v

DRIVE UNIT FOR 6th AXIS

CONTROL PANEL

PROGRAMMING UNIT

INTERFACE UNIT

NTROL UNIT

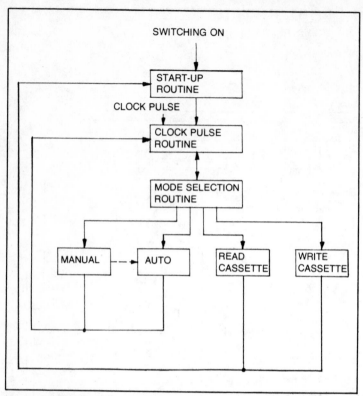

Fig. 8-14. The flow chart of the ASEA robotic system (courtesy ASEA).

A desired value signal with a frequency of 2 kHz is generated electronically in the "axis control circuit" function unit. The measuring system operates on a phase-analog principle, meaning that the phase difference between the actual value signal from the transducer and the desired value signal is directly proportional to the position error. The function unit contains a converter which converts the phase difference to an analog error signal which controls the servosystem.

During axis motion, the computer regularly reads out orders which result in phase-shifting of the desired value signal. A phase difference between the desired and actual value signals will cause the robot axis to move and therefore the resolver to rotate in order to reduce the phase difference. During motion at constant speed, the axis lags behind the commanded value with a particular lag which increases with speed. On starting and stopping, the acceleration or retardation is controlled automatically in such a way as to prevent overshoot.

The "axis control circuit" function unit contains two circuits which monitor the value of the lag. The condition of these circuits is scanned regularly and read into the computer. If the lag exceeds a certain maximum permitted value, emergency stop takes place and the robot stops. The second circuit detects whether the lag is within a few robot increments, the zero zone. During fine positioning with the robot, the next robot instruction does not begin until the axis has taken up position, that is, the lag is less than the zero zone.

Servoamplifier

The servoamplifier is part of the position control system. In IRb-6 it consists of a transistorized chopper and in IRb-60 a thyristor converter is used.

The drive units of all axes in IRb-6 are supplied by a common power supply rectifier; see Fig. 8-15. The positive or negative supply voltage is supplied to the drive unit motor circuit at a frequency of about 1 kHz. The motor voltage is controlled by varying the pulse length. Figure 8-15 shows the block diagram.

In IRb-60 the ac voltage is fed directly to the drive units after reduction by transformer as shown in Fig. 8-16. The drive unit is a conventional thyristor converter with a circulation current reactor. The current and speed are controlled in the same way as IRb-6, but in this case by controlling the firing angle of the thyristors.

In both versions there is an inductor connected in series with the motor; the purpose of this is to reduce current pulsations due to the pulsating nature of the motor drive voltage.

An analog error signal which serves as the speed reference for the drive motor is obtained from the "axis control circuit" function unit. The actual speed is measured by a tachogenerator coupled to the drive motor. The speed reference is compared with the tachogenerator voltage in the speed regulator of the drive unit. The difference is amplified

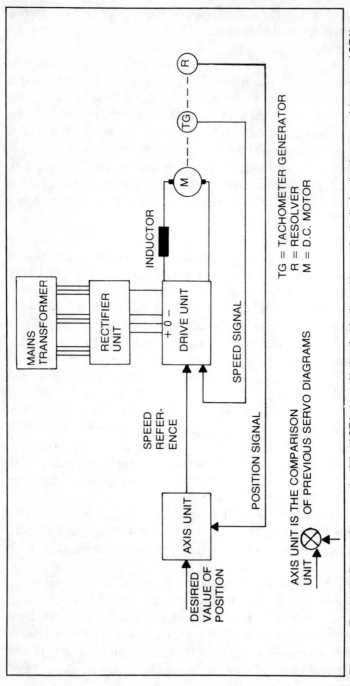

INDUCTOR

TG = TACHOMETER GENERATOR
R = RESOLVER
M = D.C. MOTOR

MAINS
TRANSFORMER

RECTIFIER
UNIT

+ 0 −

DRIVE UNIT

SPEED SIGNAL

SPEED
REFER-
ENCE

POSITION SIGNAL

AXIS UNIT

DESIRED
VALUE OF
POSITION

AXIS UNIT IS THE COMPARISON
OF PREVIOUS SERVO DIAGRAMS

Fig. 8-15. The block diagram for one axis of the ASEA robot. Notice that both position and velocity feedback are used (courtesy ASEA).

306

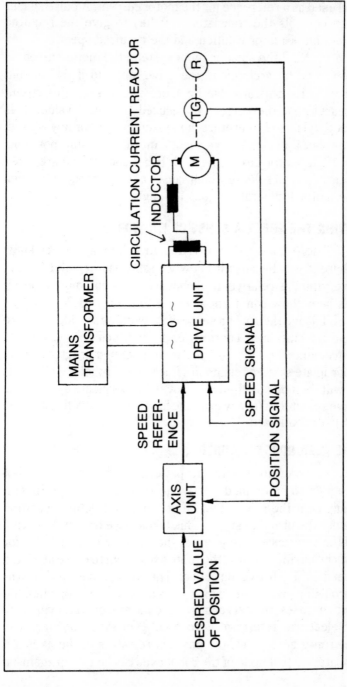

Fig. 8-16. The block diagram of the servoamplifier which uses a current reactor in the IRb-60 robot (courtesy ASEA).

307

and used as a reference for the motor current. A pulse shaper then controls the transistor switches to give the required direction of motor rotation and the required speed.

In IRb-6 the smaller robot system, the motor current is monitored by a circuit in the drive unit and if the current exceeds its critical value for more than 5 sec., this circuit ensures that the current is reduced to a safe value. This function is used to protect the system, when for any reason, the robot arm comes up against a mechanical limit position.

The motor circuits of both IRb-6 and IRb-60 are fitted with a thermal release which causes an emergency stop when the motor temperature goes too high.

GIVING THE ROBOT A SENSE OF ITS WORLD

There used to be a saying that "mobile object knew where it was because it knew where it had been." It's true! Everything is relative. If a robot arm does not know where it has been, how can it know where to move next?

To give the robot a sense of its working world, that is, to give it some base coordinates from which to measure its movements, and then to give it well defined programs in its coordinate system results in a fine state of robotic well being. It will then be able to make moves in coordinate system you have selected. We now examine how a world is defined for the ASEA robotic arm.

THE ASEA ROBOT'S WORLD

In order to enable the patterns of motion of a robot program to be defined relative to the rest of the peripheral equipment the robot must be set to a well-defined position before the program starts. This takes place on synchronizing to the reference zero point of the robot. When an order for synchronizing is given, the robot axes begin to move at a fixed speed in a positive direction. The axes move until a synchronizing switch—one for each axis—is actuated; when the switch closes, the axis continues to the nearest resolver zero, or electrical reference zero point. On synchronizing to a reference zero point, the position registers of the axes are updated so that none of the robot axes can be run up against a mechanical limit position. These position registers are stored in the memory. The working range of the axes is stored in the control program, and if any axis attempts to exceed the value, the motion is interrupted and a warning lamp lights up.

The robot can also be synchronized to its actual position. This means clearing the position error for a brief instant on the "axis control circuit" function unit. This takes place when the robot is switched on, to prevent the robot jerking when the drive motors engage, and on emergency stop, for rapid stopping of the robot.

LOSS OF MEMORY POSSIBILITY WITH MICROCOMPUTERS

It is a constantly possible horror story. Somehow the electrical supply system to the plant or robot was interrupted and the inevitable result was that the memory lost all of its information! It is not necessary that the supply voltage be lost entirely, sometimes this might happen if the supply voltage drops below a critical level. Of course there are other possibilities which might affect the robot's computer brain such as temperature extremes, static, and in case of discs, dust and dirt. In the case of a good robotic system an alternative supply is incorporated into the system. This is how the ASEA robot system handles the problem:

Since the program memory is made up of semiconductor circuits, the stored information will be lost if the +5 V supply falls below a critical value. To eliminate the necessity for reading the program from the cassette after brief supply voltage failures, the control equipment is available with standby battery supply for the program memory. The blanking plate on the control panel is then replaced by a maintenance-free battery unit. A standby supply unit must also be fitted to the control equipment. The program memory receives its +5 V supply from the battery. If the mains voltage is available, the battery is continuously charged from the standby supply unit. These are the technical data of the battery package:

●Life expectancy at least 3 years
●Capacity with maximum program memory equivalent to a mains voltage failure of at least 45 minutes

●Charging time for discharged battery, approximately 6 hours.

A red indicator lamp on the battery unit lights up if the battery is discharging. There is also a switch to disconnect the standby battery function for expected long disappearances of the mains supply voltage if, for example, the main switch is off.

If the voltages for the electronics section of the control equipment disappear or if their value becomes incorrect, either too high or too low, this is detected by a monitoring circuit, and the mains voltage to the control equipment is switched off. Only the standby power supply unit (where fitted) continues to receive mains voltage in order to maintain the battery charge.

The system includes a circuit which monitors the voltage supply of the program memory. If at any time this voltage falls below its critical value, a flip-flop is set and a warning lamp indicating loss of program lights up on the control panel when the mains voltage is restored. The program must then be re-written into the memory.

SIGNAL TRANSFERENCE IN THE ASEA ROBOTIC SYSTEM

There are tapes and recorders used to store programs and other types of data and these are conventional so we won't spend time on this aspect of the system except to note that they are used and have the normal advantages and disadvantages. We also realize that from the control rack of any robot system, a multitude of signals must pass to and from the robot and its brain section. We have discussed this to some extent in the previous chapters. Here is an actual working example system for us to examine with respect to this important area.

Signals pass between the control cabinet and the ASEA robot for the following:

Dc motors
Resolvers   From three to six
Tachogenerators   electrically-controlled
Synchronizing switches   axes

Solenoid valve	For pneumatic control of
Limit switches	tilt or turn and 6th axis
	(replaces electrical
	control)

Grippers 1 and 2, which are controlled by solenoid valves

Ac-operated grippers or alternatively, search devices with an acknowledgement signal

Connection leads to connect the programming unit to the robot

Protective earthing of robot

In order to match the control equipment for various applications it may be necessary to duplicate a number of important control functions and to control external units or detect various acknowledgements. The available signals with maximum voltage and current data are listed below. All input and output signals are filtered to suppress electrical interference. Shielded cables must be used for the input and output signals, and the leads must be connected to terminals in the cabinet. The cabinet wall has a sealed entry for the external cable. Following is a description of the external connections of this system.

Output signals, 4 contact functions

●Contact closed if control equipment is operating (220 V, 1 A)

●Contact closed when the OPERATION pushbutton is pressed (24 V, 200 mA)

●Contact open on emergency stop (220 V, 1 A)

●Contact open when the EMERGENCY STOP pushbutton is pressed (24 V, 200 mA)

Input signals, 3 contact functions

●Open contact gives emergency stop (24 V, 1 A)

●Contact closure gives "program start" (24 V, 10 mA)

●Contact closure gives "program stop" (24 V, 10 mA)

The contact closure function for "program stop" has priority over the "program start" function.

Output signal to drive relays

●14 outputs numbered 1 to 14

The relay driver stages, which are current-sinking, are designed for +24 V and 150 mA maximum. The total current in the output stages must not exceed 2 A.

The programmed outputs are used in automatic running.

Interlocking inputs for the robot program

●16 inputs numbered 1 to 16.

The input stages convert the contact functions or signals from current-sinking driver stages into logical signals suitable for the control equipment. A contact or current-sinking driver stage must be capable of connecting 10 mA to 0 V and must be dimensioned for +24 V. A 2.5 ms filter circuit is provided to eliminate the effect of contact bounce and high-frequency interference.

The programmed inputs are used in automatic running.

Gripping tools and search devices signals and connections

The standard version of the ASEA robot is supplied with electrical wiring to a socket on the robot arm for the connection of search devices or other types of grippers. Different connections must be made in the control cabinet depending on the function for which the socket is to be used. The following signals are available, depending on the application.

a. Connection of search devices

●± 15 V to supply search device electronics (max 0.5 A)

●Input to control system for the "search-stop" function. The search-stop contact or transistor must be dimensioned for + 24 V and be capable of connecting 10 mA to 0 V.

b. Connection of valves to control grippers

●2 make contacts (220 V, 2 A)

●An option unit must be added to the control cabinet.

Two solenoid valves to operate two pneumatically controlled grippers are also available as extra equipment. The valves are controlled by signals from the control unit. Compressed-air lines inside the upper arm are fitted to the robot as standard.

The procedures in checking cables for proper signals and to check these cables for broken wires and/or bad connections in sockets is the same as for any large piece of electronic equipment. There are many relays used in this unit and these must be maintained so that they provide proper connections between their moving points and the fixed points. Cleanliness, spark prevention, reduction of pitting, and heating are some areas which must be considered. Sealed relays or solid state relays tend to reduce problems in relay maintenance and in the switching of large voltages and small signal voltages in any robotics system.

CONTROLLING THE ASEA ROBOT

On the control equipment there is a control panel with controls to start the industrial robot system and to select various operating modes. For normal operation of the robot system, that is, when the robot is programmed, only these controls are required. If the equipment has standby battery supply of the program memory, the battery package is mounted in a module on the control panel.

What the Control Switches Control

Switch	Function
MAIN SWITCH	To switch the power supply on and off
a. I	Power supply connected to robot system if MAINS ON is alight.
b. 0	Power supply switched on. Content of program memory will be lost unless standby battery supply is provided.
PROGRAM SELECTION 1-4	The control system can store 4 programs in its memory. The switches are used to select the programs in automatic running and when programming a new program, or when using the tape recorder. A yellow indicator lamp lights up for the selected robot program(s).
a. OFF	This program is not used.
b. ON	This program is connected

BATTERY	The switch is on the battery unit and is used to switch the standby battery supply function on or off.
a. OFF	Standby supply of program memory from battery is switched off.
b. ON	The battery is switched on and supplies the program memory in the event of power supply failure, to prevent the content being lost.
OPERATION	Power supply connected to robot motors. Yellow indicator lamp lights up. Control system is in the operation condition.
STANDBY	Starts the control system after power supply has been switched on or after disconnection of the supply voltage for the robot motors. Yellow indicator lamp lights up. The control system is in the standby condition.
SYNCHRONIZING	A red indicator lamp lights up after the power supply has been switched *on* or after emergency stop. If the button is held pressed the robot axes go to their synchronizing position.

EMERGENCY STOP	Immediately stops all robot motions and sets the system in the standby condition. The *EMERGENCY STOP* lamp lights up.
RESET EMERGENCY STOP	This pushbutton resets the emergency stop circuit if the contact circuit which caused the emergency stop has been closed again.
OPERATING MODE	There are four pushbuttons with yellow indicator lamps for operating mode selection.
a. AUTO	Automatic operation with one or several programs to be executed consecutively in order. Yellow indicator lamp lights up.
b. MANUAL	Manual mode. The robot can be synchronized, or operated and programmed by means of the controls on the programming unit. Programs can be tested step-by-step. Yellow indicator lamp lights up.
c. READ CASSETTE	A robot program can be read in from the magnetic tape cassette.
d. WRITE CASSETTE	Transfer of a robot program from the program memory to a cassette tape. Yellow indicator lamp lights up.

PROGRAM START	In the automatic mode, the robot program starts. In the manual mode an instruction is executed whenever PROGRAM START is pressed. The pushbutton is also used to start reading from, and transfer to, tape cassette in the "read cassette" and "write cassette" modes. A green indicator lamp lights up during activity.
PROGRAM STOP	In automatic mode, stops the robot program in progress; in manual mode stops the instruction which is being executed. The *PROGRAM START* indicator lamp goes out. In the "read cassette" and "write cassette" modes the tape recorder is stopped.
POWER ON	Yellow lamp lights up if main switch is closed and power supply voltage is available.
PROGRAM LOST	Red lamp lights up to indicate that the content of the program memory has been destroyed.
EXCESS TEMPERATURE	Red lamp lights up to indicate excessively high temperature in control cabinet.
OPERATOR ERROR	Red lamp lights up for incorrect operation of controls, e.g. if a switch is set to the wrong position or if an incorrect mode is set.

EMERGENCY STOP	Red lamp lights up after an emergency stop has taken place or after internal emergency stop initiation from the control system.
BATTERY OPERATION	This lamp is situated on the battery unit and gives a red light when the battery is discharging.

TEACHING THE ASEA ROBOT

In an earlier chapter we learned how a hand-held unit can be used to make the robot arm go through the desired motions to accomplish a task. As the arm gets into various positions we press a learning button which causes that information to be placed in the computer's memory bank. In essence the arm is jogged or wiggled into the various positions it must have to do the tasks as you teach it what to do. You jog or wiggle the arm by depressing various buttons on the hand held unit. It follows your every command and the process has a fascination all its own. In a way it is like steering a radio-controlled model airplane or car or boat by slight movements of the joystick or the steering wheel. But this is a big and powerful unit which, once it has learned its job, will then, when put into automatic operation, perform that job for you almost endlessly! And do it with a constant precision and speed that always makes those concerned with budgets smile broadly! So be it. Let us see what happens when we program the ASEA robot.

PROGRAMMING THE ASEA ROBOTIC ARM

The portable programming unit which can be plugged in at the control cabinet or at the robot, is used to re-program the robot for new applications. When the robot is working in the normal operating mode the program unit can be used in conjunction with other robot systems. The robot is programmed by using the pushbuttons on the programming unit to run it to the positions it is required to take up in automatic operation. Each position is stored in the program memory when an instruction key on the programming unit is pressed. In addition to positioning instructions, it is possible, for example, to store in the program memory instructions for the operation of grippers, opening and closing of a number of outputs, testing of a number of interlock inputs, time-lag and repeating. The programmed instructions are then executed in sequence in automatic mode. The programming unit is shown in Fig. 8-17.

In order to avoid tedious reprogramming after long periods of power supply disconnection, or when more than four programs are to be used with the robot, programs can be stored on cassette tapes.

Some interesting aspects of the remote control panel are section D, which controls the arm, and E which has the control buttons for the wrist. Notice how the movements are designated on this panel. Numbers have been added to the button faces to make it easier to discuss their individual operation. Look at section D first.

Buttons one and two control a rotary movement of the arm assembly by rotating the base structure. Refer to Fig. 8-7 where the Greek symbols for the angles are shown. The control of the lower pivot of the arm is accomplished by buttons 3 and 4, and the control of the upper segment of the arm is with buttons 5 and 6. For the wrist section (E) buttons 1 and 2 control the tilting of the wrist, while the buttons 3 and 4 control the rotation of the wrist clockwise or counterclockwise.

Since the angles in Fig. 8-7 are expressed in Greek letters it might be of help to review this symbology. The following short table shows the letters used.

Greek Name	Capitol Letter	Lowercase Letter
Alpha	A	$\alpha =$
Theta	Θ	θ, ϑ
Phi	Φ	$Z = \rho = \varphi$
Upsilon	Y	υ
Tau	T	$\tau \; (\tau)$

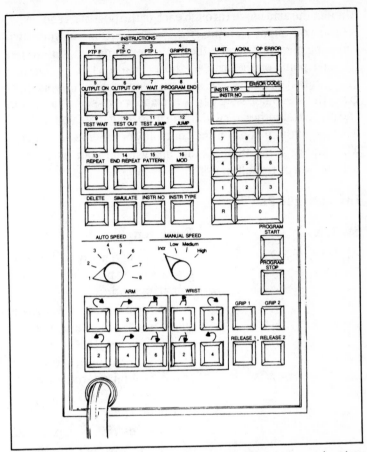

Fig. 8-17. The remote programming unit for the ASEA robotic arm (courtesy ASEA).

Now, having identified each of the pivoting joints and the directions of motion of the arm segments and the wrist itself. Let us examine what the other buttons do in the following table. Notice that you must sometimes push two or even more buttons, and set a dial switch to a proper setting in order to get the arm ready for whatever kind of commands you want to give it. It can be instructive, if you want to get a feel for programming of a robot arm such as this, to imagine that you have the remote panel before you—look at Fig. 8-17—then look at the robot arm in Fig. 8-6 and see if you can write down which buttons had to be pushed and in what sequence in order to make the arm move through each of the positions shown. Of course you might also imagine some positions of your own choosing and write down the button pushing program to get the end point of the arm to those positions. You can also imagine that grippers are attached and include in your program the steps necessary to grasp and move something from one place to another. Try it! In the following table you will find what everything does on the control panel. Some statements will require some thinking and analysis on your part, but it will be worth the effort.

THE CONTROL PANEL

Switch	Function
AUTO SPEED	Selection of speed to be programmed for positioning in automatic mode. In automatic running of program, the highest speed is limited to the speed to which this switch is set. In step-by-step operation the speed is limited to not more than speed 6. Position 8 gives maximum speed
	Posn 7 gives 75% of max speed
	Posn 6 gives 50% of max speed
	Posn 5 gives 31% of max speed
	Posn 4 gives 13% of max speed
	Posn 3 gives 5% of max speed
	Posn 2 gives 2.5% of max speed
	Posn 1 gives 1.3% of max speed
MANUAL SPEED	Speed selection for manual operation of the robot. The motion is executed by means of the ARM and WRIST pushbuttons.

a. INCR	The relevant robot axis moves one increment
b. LOW	The relevant robot axis moves at 1.3% of maximum speed
c. MEDIUM	The relevant robot axis moves at 15% of maximum speed
d. HIGH	The relevant robot axis moves at 50% of maximum speed
INSTRUCTIONS	INSTRUCTIONS is a group of pushbuttons used to program the instructions to be executed when the robot is operating automatically. On programming each instruction in the program is given a serial number. The instructions themselves are designated by a type number from 1 to 16.
a. 1 PTP F	Point-to-point operation with maximum accuracy. With arguments (special numerical values) search and vertical/horizontal running can be programmed.
b. 2 PTP C	Point-to-point operation with coarse positioning. With arguments (special numerical values) search and vertical/horizontal running can be programmed.

c. 3 PTP L   Straight-line operation such that positioning is concluded simultaneously for all axes. The travel time is stated in the form of an argument.

d. 4 GRIPPERS   The states of the grippers, i.e., "on" or "off" are stored.

e. 5 OUTPUT ON   The digital output (1-14) selected on the keyboard must be "on."

f. 6 OUTPUT OFF   The digital output (1-14) selected on the keyboard must be "off."

g. 7 WAIT   Waiting time as set on the keyboard (0.1 - 9.9 s).

h. 8 PROGRAM END   End of robot program. Restart from first instruction or jump to next program.

i. 9 TEST WAIT   Wait until the interlock input (1-16) selected on the keyboard is "on".

j. 10 TEST OUT   Jump to next program if the interlock input (1-16) selected on the keyboard is "on"; otherwise continue with next instruction.

k. 11 TEST JUMP   Jump over next instruction if the interlock input (1-16) selected on the keyboard is "on"; otherwise continue with next instruction.

l. 12 JUMP   Jump to the instruction number selected on the keyboard.

m. 13 REPEAT   Repeat the sequence the number of times keyed in on the keyboard. This instruction also marks the beginning of the sequence.

n. 14 END REPEAT   Marks the end of a repeat sequence. The system jumps to the next instruction when the sequence has been repeated the programmed number of times.

o. 15 PATTERN   Marks the beginning of a pattern and defines the common part of a number of different pattern procedures.

p. 16 MOD   Marks the beginning of each part of a pattern to be modified.

q. DELETE   For deletion of the program instruction with the number displayed.

r. SIMULATE   Used when program testing to simulate that the selected interlock input is "on."

s. INSTR NO   To modify a program instruction or start within a program. The relevant number is keyed in on the keyboard.

t. INSTR TYPE   Used to display on the keyboard the type of instruction stored at the selected program instruction.

GRIP 1	Gripper 1 closes
RELEASE 1	Gripper 1 opens.
GRIP 2	Gripper 2 closes
RELEASE 2	Gripper 2 opens

ARM

Used for manual operation of the ϕ-, θ- and α-axes of the robot. The travel speed is selected with the MANUAL SPEED switch. Several axes can be operated simultaneously.

When only one of α and θ are operated, the control system compensates to give a purely vertical or horizontal motion.

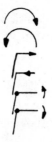

ϕ-axis rotates clockwise
ϕ-axis rotates anticlockwise
θ-axis moves forward
θ-axis moves back
α-axis moves up
α-axis moves down

WRIST

Used for manual operation of the robot wrist. Function as described for the ARM pushbuttons.

The tilt motion is compensated with the turn motion to give pure bending.

Tilt-axis moves up
Tilt-axis moves down
Turn-axis rotates clockwise
Turn-axis rotates anticlockwise

PROGRAM START	This pushbutton, which is duplicated on the control panel, is used in automatic operation to start the robot program. In manual operation, new programs can be tested step-by-step, since one instruction is performed every time the pushbutton is pressed. A green indicator lamp lights up in the pushbutton while a program or instruction is in progress.
PROGRAM STOP	This pushbutton, which is duplicated on the control panel is used to stop programs or instructions in progress; the PROGRAM START indicator lamp goes out.

Digital Keyboard	**Function**
0, 1, 2, 3, 4, 5, 6, 7, 8, 9	Used to key in numerical values and arguments belonging to the relevant instruction and to select the instruction for indication or deletion. (argument = angle)
R	Clearing before input; the numerical display panel is also cleared.

Indicator Lamps	Function
LIMIT	A red lamp lights up if, in manual mode, an attempt is made to move any axis outside its working range.
ACKNL	A green lamp lights up when a new instruction has been registered in a robot program.
OP ERROR	This red lamp lights up if any of the controls on the programming unit or control panel is wrongly operated. An error code appears on the numerical display panel.

Numerical Display	Function
ERROR CODE INSTR TYPE INSTR NO	Four-digit readout to indicate instruction number, instruction type or error code. The format is explained by the text above the display panel.

Notice that the designers of this robotic system have used indicator lights to help prevent errors. This is especially true in the area of operator error. Notice that not only does a red warning light light up, but also an error code showing what is wrong with the command or sequence appears on the numerical display panel. When one sets in all the necessary commands correctly, the whole panel is lighted up with green (go ahead) lights.

THE GRIPPERS

We need to spend some time examining the gripper operation as it applies to this robot and the others which we have discussed. We already have some knowledge of how the grippers appear and function from previous illustrations, but here we will look at these units in more detail.

The first types of end units are those that are affixed to the robot's wrist by bolts or other type locking mechanisms. Some of these types of end units are the welding torch, the paint atomizer or spraying unit, drilling mechanisms, and perhaps even nut tightening units that can tighten to the correct torque, precisely. There are many other end units which fit into this category as you can imagine.

The second type of gripper is that which may have a pincers arrangement or claw using two or more fingers that can grip and hold various objects while the arm either moves to new positions, or adjusts the object held to another mechanism which may polish, sand, de-burr, paint, or what-ever. We have noticed that many arrangements of the claw or pincers type end-unit are possible, one of which is illustrated in Fig. 8-18.

One interesting aspect of this gripper is the lever ar-rangement which makes the jaws open and close as the drive mechanism pulls or pushes the levers at point A. A second item of interest in this unit is that the jaws have a soft face which permits some compliance in the grasping action. What this means is that the jaw faces can conform to the shapes of some objects to grasp them better, and are not rigid as would be the case if the jaw faces were solid metal. Also, when you have this compliance capability, it is an easy step to incorpo-rate some pressure activated switches in the complying material such that when the object is grasped a switch is closed that lets the computer know that the object is being held. The computer can then order, perhaps, a slightly in-creased tightening of the jaws. Not enough to damage any-thing, but enough to insure that the object won't slide out of the gripper when the robot arm moves it. In systems where the jaws simply close to a certain dimension and are stopped, if the size of the object changes or its physical shape changes, then the object may not be properly grasped at all. One has to re-program the closing dimension in order for the gripper to close satisfactorily on the new object's shape or configura-tion. Compliance is a good thing and we can think of only one

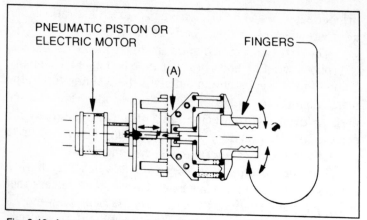

PNEUMATIC PISTON OR ELECTRIC MOTOR

FINGERS

(A)

Fig. 8-18. A pneumatic type gripper may also be electrically operated. In the latter case the pneumatic piston is replaced with an electric motor. Here we see a diagram which shows the mechanics of operation of one type gripper. Notice that the "fingers" open in a rotary movement, not a parallel, translatory movement.

possible disadvantage. The material may wear out and have to be replaced.

Examine Fig. 8-19 where two types of gripper arrangements are shown. Manca, Inc. manufactures various types of grippers for robot machines. Notice that in A two fingers which are pivoted can be caused to move through an angular rotation if they are connected by a suitable screw and nut-type linkage to a motor's gear train shaft. As the motor rotates the shaft screw it will pull or push the linkages to cause them to open or close, as shown in B where the position is closed. The illustration in B of Fig. 8-19 also shows the disadvantage of this type gripper action. Because the motion is angular the tips of the fingers tend to come together while the body of the finger has a rather wide gap or spacing. Of course, as shown by the dotted lines in B, one might use a compliant material which can compress and this helps to overcome this difficulty.

If we examine further in Fig. 8-19 we notice that part C will overcome this disadvantage of angular motion because here the electric motor will cause the fingers to come together in a parallel motion instead of the angular motion just described. In part D of the illustration we see how nicely the fingers close together. This may be an ideal method for palletizing, or similar type work for the robot.

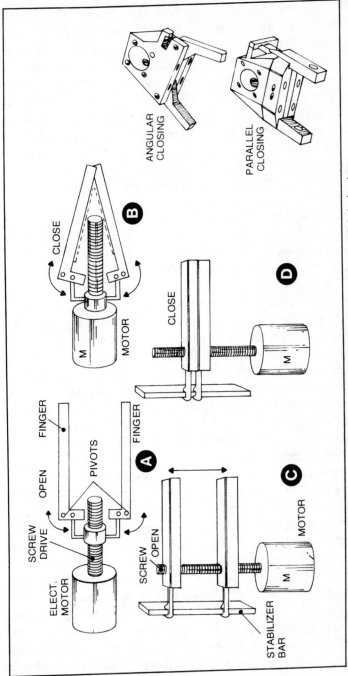

Fig. 8-19. Some types of grippers, and the operation of an angular closing type and a parallel closing type.

329

Only two fingers have been shown. Please realize that in some cases it may be necessary to use three or more fingers and these can be made a part of a somewhat elaborate mechanical system which can be activated by a motor arrangement without undue difficulty. The principles of operation may be the same as we have herein illustrated. In the pictorial we show two such units courtesy of Manca, Inc. In some applications where a slight opening and closing of the fingers is all that is necessary to grip the object the rotary or angular system might be very satisfactory. Since it closes only slightly, you would not have the finger-tip condition we have shown at B or Fig. 8-19.

When one has fingers which can move in a parallel manner, it is mandatory that guide rods be used to hold the fingers in line while the motorized section opens or closes them. We have called this part of the gripper a stabilizer bar in Fig. 8-19.

Now we turn our attention to the fingers portion of the grippers. We use Fig. 8-20 as an aid to our thinking. The fingers may be just bar-like metal units or they may be flat surfaces of metal. They may have some compliance or sticky material attached to the flat metallic surface as shown in B, or they may have an indent of some particular shape as shown in C. The gripper's fingers will be designed and shaped according to the type objects they are to handle and may include even pneumatic suction cups on the inside edges which will afford a tighter grip or better holding capability on fragile or light weight objects which must be moved. In such a case a very delicate closing of the gripper can be accommodated so that it doesn't close tightly at all. The gripper's fingers will close enough to bring the suction cups against the object and then they will stop moving. The suction of the cups then will hold the object while it is being moved by the robot arm.

Figure 8-21 is a typical commercial, angular type gripper for an industrial robot. This unit can be operated by a hydraulic system, a pneumatic system, or a spring system. Notice the limit switch labeled here as a proximity switch. Realize that with hydraulic or pneumatic systems the backpressure in these systems can cause the proximity

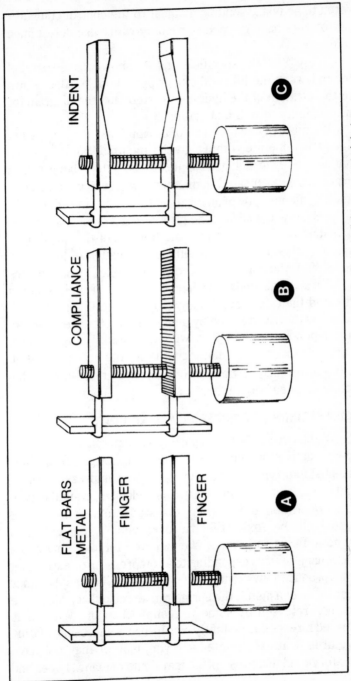

Fig. 8-20. Some finger considerations on grippers. Note that each one shown here is a parallel type operated device.

331

switch to activate, sending a signal to the control computer that the jaws have gripped something with a pre-determined force.

In Fig. 8-22 the specifications are shown for the parallel motion jaws of the Fibro-Manta gripper module. Also shown are the diagram of the hydraulic system and the diagram for the proximity switch installation.

We normally do not think of a gripper as being some unit which must operate within a given time limit, but it is such a unit, especially in industrial applications. If the robot is to be a part of a large automated chain, then it will receive the units it handles in some pre-planned time sequence. In order to do this precisely, the jaws of the gripper must open at a particular instant of time, close in some fraction of time, and open again on schedule. Note the time specification of the unit in Fig. 8-22. If the unit is operated by hydraulics or pneumatics the time is 0.2 seconds, which is very fast. If the unit is spring activated then the time is longer by 50% to 0.3 seconds. If these jaws were operated by an electric motor the time will vary depending on how they were operated. Using a clutch, gears, or a direct drive motor could vary the time to substantially more than the fraction of a second stated for the hydraulic or pneumatic units.

A COMPARISON OF GRIPPER POWERING UNITS

We can make some comparisons among the three or four types of activation units used to power robot grippers. First let us list four types. Hydraulic uses a noncompressible oil or similar working fluid. Pneumatic uses dry air or similar type gas for operation. If the gas is exhausted into the atmosphere it must not be toxic. Electric uses electric currents and magnetic fields to produce the required forces. Spring is a reactionary device responding to a compressing or expanding force produced by some other means. Like the hammer mechanism of a pistol, the spring can be cocked at a slow rate and then released by some trigger mechanism. When it is released it responds violently, and in a very short span of time it returns to its rest state. A spring may or may not have vibratory tendencies. A spring may come in many shapes and

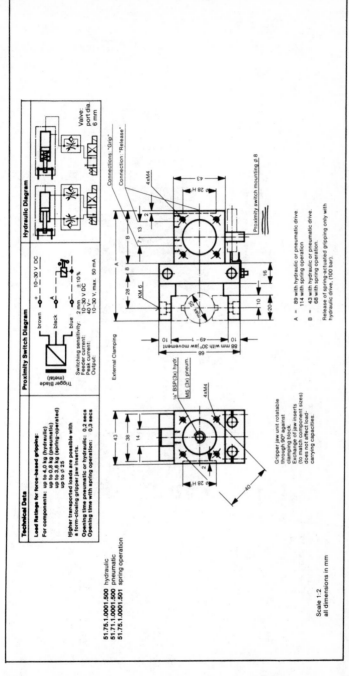

Fig. 8-21. A commercial, angular closing type gripper, and specifications (courtesy Fibro-Manta).

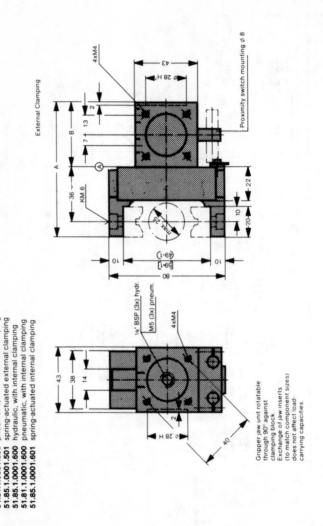

External Clamping

51.85.1.0001.500   hydraulic, with external clamping
51.81.1.0001.500   pneumatic, with external clamping
51.85.1.0001.501   spring-actuated external clamping
51.85.1.0001.600   hydraulic, with internal clamping
51.81.1.0001.600   pneumatic, with internal clamping
51.85.1.0001.601   spring-actuated internal clamping

Proximity switch mounting ⌀ 8

4xM4
⌀ 28 H7
43
2
13
7
B
A
36
KM 6
max 25
22
10
20
10
59±1
49-1
10
80

⅛" BSP (3x) hydr.
M5 (3x) pneum.
4xM4
43
38
14
⌀ 28 H
2
40

Gripper jaw unit rotatable
through 90° against
clamping block.
Exchange of jaw inserts
(to match component sizes)
does not affect load-
carrying capacities

334

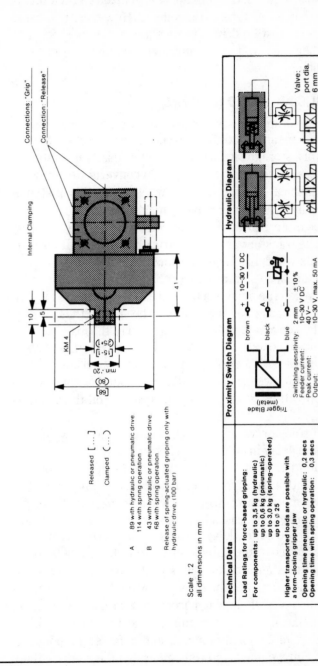

Fig. 8-22. A commercial closing, parallel type gripper. This may be electric, hydraulic, or pneumatically operated. The hydraulic diagram is shown at bottom right (courtesy Fibro-Manta).

335

forms and sizes from the commonly visualized circular unit to a simple bar-lever type arrangement. However, it is reactionary, and must be driven to its state of energy storage by some other means. This requirement may or may not be a disadvantage.

HYDRAULIC-ACTIVATED GRIPPERS

The hydraulic system works with a non-compressible fluid. This means that there is no give in the system. It consists of a storage tank a pump or compressed air system to apply pressure to this oil. The piston converts the pressure into mechanical motion. The flow of oil to this piston is regulated by the transfer valve, which is the means by which the robot brain can control the movement of the arm segments. The transfer valve is electrically operated. It consists of a spindle which has large and small sections machined into it. When the large sections cover the oil-flow ports, the oil is shut off. When the small sections are in line with the oil-flow ports, the oil can get by the spindle. The spindle thus must be centered when a neutral signal is applied to the magnetic coil which encompasses the spindle. The centering is usually accomplished with a spring which is factory adjusted. From the description we can assume that the speed of operation of the piston will be governed by the speed of oil flow, and its pressure.

The transfer valve is subject to two forces which can cause some delay in operation. They are stiction and friction, the friction being of a viscous type, while the stiction is a molecular attraction which tends to cause a kind of binding when two metal parts are very closely machined and put together. In other words, a larger force would be required and there would be some delay if the spindle were at rest when the signal to move the spindle energized the magnetic field. To overcome this problem, it is customary to apply a low frequency ac current to the magnetic winding and this causes the spindle to vibrate at a high frequency with low amplitude. Since it is thus in constant motion around its neutral position, the stiction force is minimized, and a faster response is obtained. But we must also remember that the working fluid, oil, has a high density, and thus, even though it is light-weight, (low SAE number) it also adds delay in motion. Thus, while a hydraulic system is very fast in operation, it does have some delay forces acting on its elements.

We can summarize the hydraulic type gripper, then, by saying that it is very powerful and very fast. It requires some ancillary equipment and it requires high pressure secure lines, fittings, and seals.

THE PNEUMATIC SYSTEM

The air or pneumatic system is operated with compressed air or a gas of similar type. Air is compressible, thus under large back forces the system must apply more pressure to adjust or the system will "give" a little. If the additional pressure is applied and the back force is reduced suddenly, the system must accommodate and adjust, and so we have some tendency toward an oscillatory motion of the output member of the system under these conditions. But air is much less dense than oil and thus can move faster and so the pneumatic system is probably the fastest system possible. Valves, seals, and the working gas itself must be kept clean and dry which means filters and other ancillary equipment to do this task. The air, or inert gas, may be exhausted into the atmosphere and the storage tank may be supplied from the atmosphere using a pump, so, unless the system works in a vacuum or under water it can get its working fluid from the atmosphere and put it back into the atmosphere after using it! The pneumatic system is fast and has the happy advantage of being usable in reverse, or in a suction capacity. It can hold items and can attract them and be otherwise of value in a robotic system.

Because the air is compressible, it can be stored in tanks where it can be drawn in low amounts for a long time to activate some pistons through a transfer valve system similar to that used in the hydraulic applications. Compressed air can be used to pressurize a hydraulic system, and if the air is contained in tanks, then the hydraulic system can be operated without a pump or storage tank separate from the air-oil supply cylinder. Of course in this arrangement the oil is not returned to the cylinder after use, it must be expended out of the system in some way.

The seals in an air operated system must be tight because the density of air is less than oil. A transfer valve used in the air system must also overcome stiction and viscous friction even though the viscosity of the air is very low. Then, as we have stated, the air must be dry (inert gas is often used because of this), and the air must be clean, requiring filters with micron size holes in them.

THE ELECTRICALLY OPERATED GRIPPER

Perhaps the most versatile powering system is electricity. It is easily controlled through well known means, and as we have seen, motors can be mounted directly at the point of usage and screw drives or clutches or cables can be used to transmit the mechanical power. Electric motors are reliable, have long life, can be well sealed, have no leakage problems, require no tight seals and so on.

The motor also has a disadvantage or two. First, in order to get lots of power one must have a powerful motor which means bulk in size, and they are heavy. If one uses a gear train to develop the required power, delays in movement must be expected unless the motor turns at a fantastic speed. The armature is heavy so there is always the inertial overshoot problem to content with, snap-close clutches tend to wear and have to be replaced and cable drive systems can become loose if they do not receive constant attention. All of these problems can be overcome, however, and the advantages make it worth it.

Some advantages of additional importance are that the electric motors can be controlled with either ac or dc electricity. They can develop, through their own back emf, voltages and currents proportional to their speed and these can be read directly for control purposes. This prevents or helps to prevent overshoot and subsequent oscillation or hunting. Electric stepper motors can be moved precisely with electronic pulses, and a computer can keep track of exactly how much the motor has moved. Solenoid-type magnetic units can cock spring loaded units, which then can be released by another magnetic trigger creating very fast operation.

SUMMARY ON GRIPPER POWERING METHODS

To summarize on the powering system for grippers we might list the pneumatic system as fastest, using compressible gas, requiring excellent filtering and tight seals, and producing a good powering level. It can be used on delicate operations.

The hydraulic system is the most powerful, requires ancillary equipment to pressurize the oil, needs good tight seals and is slightly slower than the pneumatic system.

The electric system is versatile and comes in many forms, is easy to control, but can be bulky and heavy if lots of power is required. It can also be used on delicate operations.

GRIPPER SENSORS

We must not leave this discussion concerning grippers without mentioning the feedback units which can be used with such end-devices. We know that many countries are experimenting with fibre-optic systems built into the gripper to give the control computer some knowledge of the proximity of objects, the shape of objects, and when the gripper has such items in its grasp.

Examine Fig. 8-23 where the angular type gripper might not be most suitable because the light beams might not constantly impinge on the receptor units in the fingers. In A we see the static condition of a gripper and the light beams from the generators going directly across the open space to the receptors where they cause signals to be sent to the computer. If you will imagine the top finger and bottom finger pivoting as they open to grasp something, you will see how the light beams will be moved toward the end of the finger and no longer impinge perpendicularly on the lower finger receptors. See part B of Fig. 8-23. You might consider this to be a valuable effect if you let the beams activate the receptors, say 2, 3, 4, and then 3, 4 and then finally, just 4 and let the signals thereby generated inform the computer controller just how far the jaws have opened. Then if the beam is interrupted completely, the computer knows an object has come between the fingers and can initiate closing of them.

If we consider a parallel gripper finger movement such as previously described, then we note that all beams stay positioned on their respective receptors from wide open to close position of the fingers. This fact can be used to let the computer know how far inside the jaws an object has been positioned as well as the fact that it has been positioned therein. This could be of value in some applications.

Some applications might use a gripper-finger system wherein light would be emitted from each finger outward and a receptor near the emitter would then receive a reflection when an object was encountered. This idea could be used to help the gripper find an object on a belt, or help it be re-positioned by the computer brain of the robot when items come along some feeder system with some discrepancies in their position. There are many ways in which this type of sensor might be used in a robotic system gripper.

Another sensor which has been used is the proximity switch system or the back pressure switch system which enables the robot's brain to apply just the right pressure to the fingers to firmly grasp an object once it has found it and has it within its fingers or jaws. We have had some examples of this type switch system in the Mini-Mover-5 system. The back pressure switch outlet is shown on the Fibro-Manta specification drawings, Figs. 8-21 and 8-22. We know it is relatively easy to incorporate such a switch in the fingers themselves. It can be closed when a given pressure is applied. These switch sensors are very reliable and very commonly used.

If the objects being handled are metal of a ferro-magnetic type, then one might imagine a gripper made of some non-metallic substance and inside this gripper a sensor which can generate a magnetic field which will be routed with more intensity through its windings when ferrous material comes into its field. The closer the object is to the generated magnetic field the stronger the sensor signal would be, thus we have a proximity sensing device with no moving parts. It can

340

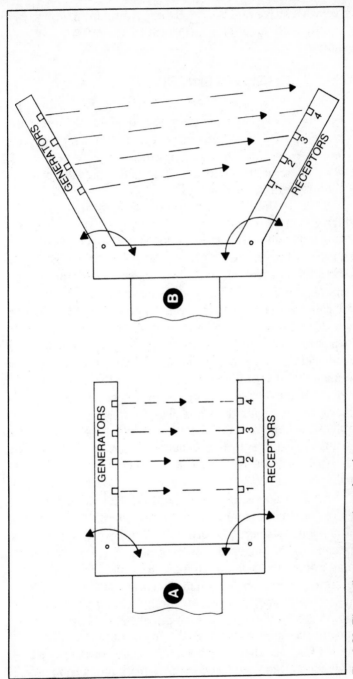

Fig. 8-23. Fiber-optic sensors systems for grippers.

be very sensitive and very reliable. There are many ways a magnetic field might be used as a sensor system in a robot's grippers.

POSITION SENSORS FOR GRIPPERS

The type sensors we have discussed have been of a proximity type. They can inform the computer that an object is present, where it is with respect to the grippers jaws, and so on. Now we want to consider some simple type sensors which can inform the computer of the actual physical position of the grippers jaws. Refer to Fig. 8-24.

At A on the figure we can see a linear potentiometer which has been connected mechanically to the bolt-screw arrangment which translates back and forth to open and close the gripper fingers. At this particular point of the mechanism the motion is a translation and so a *linear* translatory potentiometer can be used. The translatory potentiometer can be calibrated so that a fraction of a volt represents a fraction of an inch. It will give a signal amplitude directly proportional to the position of the fingers since they move together and in the same opening or closing direction. Notice that only one need be used because of the simultaneous and twin action, of the fingers. If the linear potentiometer is used in a bridge as shown to the right, then opening the fingers will (or can) produce a negative signal with respect to ground, and a closing of the fingers will produce a positive voltage. Now we have both amount of motion and direction specified by an analog voltage.

At B on that same section of Fig. 8-24 we show how a regular circular or semicircular potentiometer might be used in the same manner. The arm of the potentiometer must be fastened to the finger at the pivot point so when the finger rotates about this point it will turn the wiper. The body of the potentiometer must be fixed in place as had to be done for the linear potentiometer. The housing shell or frame on which the fingers are mounted can be used for this purpose. It would work in the same manner as the linear pot.

At C on the figure we see how a resolver might be used to give amplitude and direction information to a computer. It

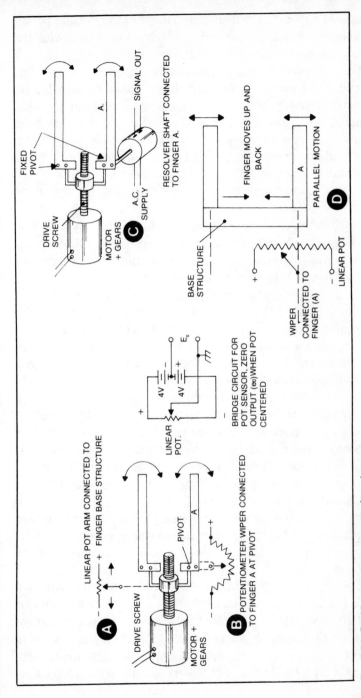

Fig. 8-24. Position sensors concepts for grippers.

343

has to be mounted in a similar manner to the circular potentiometer. It has an armature which must rotate with the movement of the finger. The stator part of this unit has to be fixed. When the rotor or armature is turned, an ac signal is produced from this unit which has a phase relationship directly proportional to the displacement. The amplitude of the signal is proportional to the amount the armature has been turned. This is an ac system which is quite commonly used in servomechanisms for position feedback information. Again, note that only one unit is required to give information on the fingers because they move simultaneously, and always in the same direction.

Finally, examine Fig. 8-24 D which is simply a parallel type gripper equipped with a linear translatory potentiometer to measure and generate signals proportional to the finger and direction. It works the same way A did.

It is possible that one might ask why we need position information on the fingers of a gripper. First because the gripper fingers want to be opened to receive or grasp something and the robot's brain will want to know that the fingers have opened to the proper degree. Second, the fingers will want to be closed only a specified amount in some cases to receive specially constructed items. Again the robot's brain wants to know that when it initiates a signal for the fingers to close, they have done so, and to the amount specified. It is much the same as our human requirements to know our hands have our fingers open to pick up something, or that our fingers are pre-positioned to make grasping of special objects easier. We also like to know when our fingers have opened to release the object, although we are not really conscious that we have obtained this feedback information. Our eyes simply tell us we have put a book on a shelf, replaced a mechanics tool, or laid down a pencil or whatever.

So we have proximity sensors and position sensors. Do we need any others? You might well answer yes without fear of contradiction. If the robot is working in an environment where temperature is a factor on the parts it handles, then, perhaps, it needs a temperature sensor to tell it when it can handle a part and when it cannot handle that part. If a grinding operation is being done then the robotic brain might like to know when the part is reaching a too-high temperature due to the abrasion, and thus pull it away from the grinding wheel. Temperature sensors in the form of thermocouples are commonly used in automobiles and in our homes on water-heaters and perhaps in other places. They can generate a voltage which can cause a switch to open or close and thus produce the high temperature warning signal. It is possible, also, that the thermocouple can generate voltages proportional to the temperature so one could have gradual control over a heat-application process if this was necessary.

What else might we use? How about a vibration sensor? This might let the robotic brain know when some object being handled was not going through its process smoothly for some reason. If the mating of a part is not a good smooth operation then excessive vibration could occur which would be felt. The vibration sensor, in turn, would sense this and send the signal alarm. Here, again, it is an item which is commonly used in the market place. This type sensor is used on large window glass to signal breakage, thus security systems have them, and instructions as to their use, and security people have experience in their use. You wouldn't want such a sensor to be too sensitive as there is always some vibratory action in a mechanical robot.

If you are an engineer or experienced experimenter then perhaps you desire some kind of speed-of-operation or speed-of-motion signal. This signal can be obtained using derivative electronic circuits, into which a part of the position signal is fed. One could also use some kind of small tachometer attached, to moving elements to produce a signal proportional to the speed of movement. Remember that the rate-of-change of position is velocity, and so a first derivative circuit whose input is the feedback position signal should produce a signal which is proportional to the velocity of movement.

SUMMARY ON GRIPPER SENSORS

There is no doubt but what a robot's gripper must have some kind of sensing devices built-in, if it is to do expanded and complex tasks. It must also have some kind of compliance so that the gripper can adjust to the shape of the object it handles, especially if the object changes in size and/or shape in various batch units. Grippers must be matched to the task they are involved with, such as high temperatures or very tight gripping.

You may be thinking ahead and wondering just how optics, or the fact that robots can be equipped with eyes, might affect the gripper sensor concept. Giving the robot eyes will assist it, but gripper sensors still may be required.

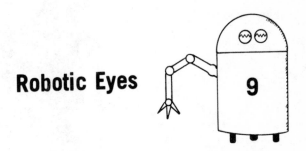

Robotic Eyes

In Chapter 4 we looked at the Cincinnati Milacron T^3 robot. Now we want to consider this robot again with respect to some features which are pertinent to what we have been discussing. As a stepping off point for this chapter, let us examine Fig. 9-1 which is another view of this modern robotic arm for industrial type applications.

We should be pretty knowledgeable about what we are looking at. There is a base section, lower arm section, its rotary electrohydraulic driving motor, the upper arm section which is operated by a piston type linear hydraulic system, and finally out to the wrist and gripper which are also operated by electrohydraulic units. Notice that in the illustration the grippers are of the parallel movement type, that the jaws are shaped to accommodate round or similar shaped objects, and that the grippers have a small thin compliance sheet built into them.

In an industrial application, of course, there can be a lot of dirt from the products, or processing of the products. In these locations and situations, of course, you would want the robot to be adequately protected by sealing, and with covers on the moving mechanisms so it won't acquire any debris which might cause premature wear, inaccurate operation, or

Fig. 9-1. The Cincinnati Milacron T³ robotic arm (courtesy Cincinnati Milacron).

excessive down time of the system. As you notice, the T³ is so protected. It is also of interest to see the complexity of the wrist giving it the equivalent of a human's wrist movements.

In the specifications for this robot, we find that the servomechanisms use both the position feedback we have discussed and also a velocity feedback. The position feedback is through the use of resolvers which operate on ac, and a tachometer-generator which produces a voltage proportional to the speed of movement. These two feedback signals enable the servocomputer to adjust the movement and operation of every axis of the robotic arm to a smooth and fast operation. You might want to go back and review Chapter 4 for the operation and teaching method employed with this robot.

We have mentioned the velocity-of-movement feedback, but perhaps we have not specified just how fast such a robotic arm moves when it does a task for us. This Milacron robotic arm can move from one inch-per-second to as fast as 127.5 inches-per-second and this velocity range can be adjusted in increments of 0.5 inch-per-second units. Its precision-in-location ability means that it can return to the same over and over with an accuracy of plus-minus .025 inch. A way of visualizing this accuracy is to know that it represents 1/40 of an inch. You know what a 32nd of an inch is, so tighten up a little on this distance and you'll have the 1/40th of an inch distance.

There are some types of applications where tolerances must be even more precise. In fitting a bolt or a shaft into a tight tolerance hole, an accuracy of 1/64th inch or even better might be desired! Even then some wobble may be incorporated into the robot's wrist movement to cause the tight fitting object to go into the hole as it is supposed to. Figure 9-2

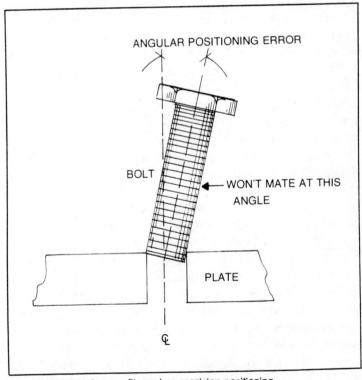

Fig. 9-2. A tight tolerance fit requires precision positioning.

illustrates how just a small off-set can make it difficult if not impossible, to get a bolt or shaft into a closely machined hole. Perhaps the problem of getting a servomechanism to have even greater than usual position accuracy on a repeatable basis will be solved, and the limit of such accuracy may astonish all of us!

When we consider an industrial robot as shown in Fig. 9-1 we also want to be aware that this type unit can be programmed to perform a series of operations with one tool as its end unit and then change to another tool to do some other operation. The wrist connection must be such that a quick disconnect and quick connection can be accommodated, and with all the power connections necessary to make the new tool operate in whatever manner it is designed to operate. When we think about this for a moment we realize that there is more to a tool changing operation that is first apparent.

One other item of information which we now should be aware of and which we have not mentioned prior to this is the concept of a rotary hydraulic actuator. They exist and they are rather common. They are much used in marine operations. They can be precisely controlled to rotate small or large amounts and they have considerable power. If you are interested in going into this subject deeper, I suggest a good text on hydraulic engineering from your local library.

It is always interesting to see a robot arm in action. If you examine Fig. 9-3 you can get one idea of how this Milacron robot appears on location and in action. One other view of this robot in action can be seen in Fig. 9-4. What is interesting about this picture is that this is just one station of a two station operation. Shown is a refrigerator liner which is made of thin gauge plastic and due to its large size is difficult to handle. One robot arm removes the liner from a moving conveyor and places it in a trim press, then it is removed from this trimming press by a second robot which again hangs it on the moving conveyor overhead. The two robots must work together in coordination and must know all about the orientation of the linear to insure it will be properly trimmed and will not be fed to the next station up-side-down!

Fig. 9-3. Cincinnati Milacron robotic arm in action in a Fort Worth, Texas plant, drilling F-16 panels (courtesy Cincinnati Milacron).

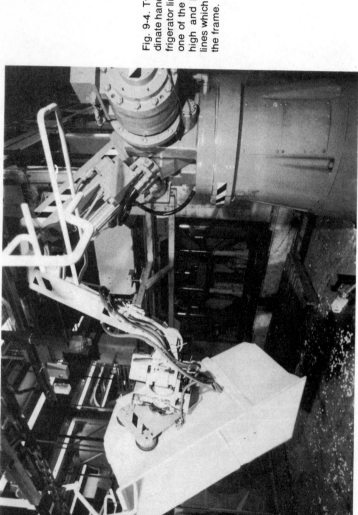

Fig. 9-4. Two Milacron robots coordinate handling the movement of refrigerator liners, but here we see just one of the robotic arms. Notice the high and low pressure hydraulic lines which are like garden hoses on the frame.

THE QUEST FOR IMPROVEMENT OF INDUSTRIAL ROBOTS

Time marches on and improvements and breakthroughs in technology make possible concepts, ideas, and things which are almost incredulous. Improving an industrial robot may not be an incredulous feat, but it is worth some time and attention. And, like our mini and micro computers, robotic systems do improve as time advances. We will examine one such development.

Turn to Fig. 9-5 where the newest of the Cincinnati Milacron robots is shown. The immediate focus of our attention is on the new wrist development shown. This is a three-roll wrist and if we look back to Fig. 9-1 we can see how the wrist complexity has been reduced in this later version. The upper arm and wrist are more slender and the powering units for the wrist and gripper are now mounted at the elbow joint of the upper arm. Physical linkages are contained inside the upper arm so that it is very clean in appearance.

It is of some importance to discuss the use of such a robot on an automobile assembly line such as at the General Motors Lakewood, Georgia plant. The robots are used to spot weld about 9600 welds per hour on as many as 48 car bodies which

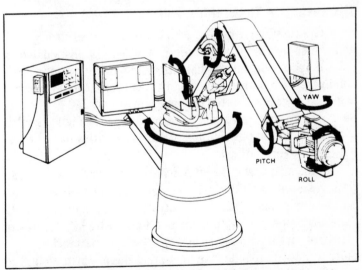

Fig. 9-5. The T^{3}R^3 Milacron robotic arm has a three-roll wrist. Compare this arm construction to that of Figure 9-1 (courtesy Cincinnati Milacron).

pass its station each hour. 200 welds per body are accomplished, and that means the robotic arm must operate quickly and precisely to do this job. Four such robotic arms are used. With a capability called tracking the robotic arm can move as the car body moves so that the welding takes place in a continuous fashion, and the conveyor line never has to be stopped. It is said that it makes no difference whether the line speeds up or slows down, the ability of the arm to operate faster or slower, going to the precise positions where the welds are to be made, is accomplished by position sensors which are attached to the conveyor belt and electrically interfaced with the robot's computer control center. Thus we learn even more about how such a robot is used and how it operates in the industrial application. With these position sensors on the conveyor belt, a continuous signal is sent to the robot's brain telling it the exact position at all times of the car bodies. Inside the robot's computer brain is a section which can take this feedback signal and use it to the previously taught welding program and the arm's position so that it moves to new locations at the proper speeds to compensate for the car's movement.

THE MILACRON THREE - ROLL WRIST

The new wrist design which we examined in Fig. 9-5, and which is called the three-roll wrist, is shaped like a sphere about 8 inches in diameter and has built into it three roll axes. Two of these axes work together to give the robot arm 230 degrees of pitch (up and down) and 230 degrees of yaw (side to side) motion. The third roll axis is aligned so that it permits continuous rotation of the wrist with whatever tool may be affixed thereto.

Once again, with remote drives to make the wrist function, resolver position feedbacks and tachometer feedbacks on each axis, and each axis controlled independently by its own servomechanism, the arm functions with a high degree of precision. Its positional accuracy is within .020 inches. It is quite an accomplishment to move that mighty arm around so quickly and yet to within that small of an error of a defined point in three dimensional space. Please note that many of the industrial robots achieve this accuracy, or better accuracies than this. The discussion of accuracy is in connection with the generic robotic generation of our time, not any single system, although we are examining the Milacron as one example of a fine robotic arm.

THE ROBOTIC EYE OR OPTICS CONTROL

I was delighted when I first saw the small robot mounted on a four wheeled cart that was like a child's wagon. It was self propelled and steerable, and carryed the robot (shaped something like the upper half of a human torso) along with astonishing speed. The head turning to look at various scenes around this robot. I was told that this was an inspection robot which was patrolling the yard of an industrial complex. It moved fast and saw everything and then relayed, through a special microwave or ultra high frequency channel, the information about what it was seeing back to a central control station where the guiding mechanism—a human being—was involved with steering, stopping, starting, and evaluating the picture received from this remotely controlled robot. Figure 9-6 illustrates the general concept of this unit.

The robot has no hands or arms. Its purpose is simply to run around the area's complex looking at everything and relaying the information back to control-central. It worked much faster on patrol than a human might, and in case there was danger from penetrators, its metal armor could shield it while it informed its master what was going on. It never tired, and aside from a pit stop now and then for a new battery supply, it could patrol constantly and for indefinitely long periods of time. Its eye protruding from the front of the round head actually is the snout of a TV camera and on its head protrude two antenna, one for transmission and the other for reception of the required TV and guidance signals. Some plaything!

But it has its purpose and can do a job for its owner. It never tries to fight back if it runs into culprits, so it doesn't need guns and it may not even need a voice although it would be an easy task to incorporate a speaker which would relay its owners voice if questioning or demanding were in order.

Let us now proceed to investigate some hard facts about just what robotic eyes are, how they work, and what their capabilities and limitations are.

There is no way at present to provide a robot with all the visual capabilities that a human has developed. To elaborate on this idea, think of all the various scenes that you can immediately identify. Think also how the human brain can identify size and shape and coloring and structure and composition and distances. Can a robot's eyes do this? Can it do it for all the countless pictures which the human brain has in its storage bins?

What we are saying is that perhaps the hobbyist's dream of having a robot, with eyes roll down the street and say good morning to Mrs. Smith and Mrs. Jones and Candy and Jim as the robot sees them and identifies them, seems unlikely. Of course, if we provide the robot's brain with pictures of these people and it can then compare the picture it sees with what we have given it, it might. Polaroid and Bell Labs have both been working on computer recognition.

In the industrial situation we find it very desirable to give a robot some eyes. If it can identify a particular part amongst many parts passing by on a conveyor belt, that can be most useful. If it can watch and follow and adjust for parts misorientation and placement and such, that also can be most advantageous. So there are many prominent companies engaged in developing robotic eyes for use in the industrial scene as we shall see. For the hobby scene, it may be some time before the use of eyes on a robot becomes worthwhile to incorporate. The basic question is what advantage is there to giving the robot some vision?

The remote-controlled robot might be clad in asbestos armor for going into a flaming floor of a home or office or warehouse and seeing what is there, and perhaps accomplishing a rescue attempt. For underwater robots, one must certainly have some vision transmitted to the control center so that a human might send commands to make the robot fix or adjust things underwater. The seeing robot can be very useful if it extends our own vision capabilities. Visual feedback may also help to eliminate the stringent accuracy requirements in

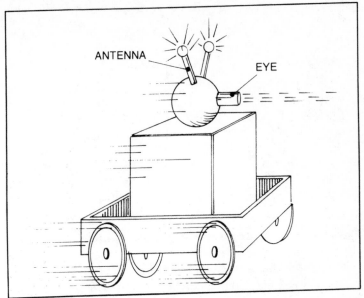

Fig. 9-6. A remote-controlled "seeing eye" robot.

the workplace. If the robot can see what it is doing, the arm can adjust the manipulator so that the task is accomplished. Seeing as we use the word here, may not be precisely the same as seeing related to the human optic operation.

Let us examine what is meant by seeing in an industrial application. A robot is going to be able to reproduce what it sees in a digitized form so it can compare it to a previously saved picture and make adjustments in the arm. Examine Fig. 9-7.

The camera is pointed where the object is to appear and instructed to follow, or keep track of it. When a picture is obtained, it must be analyzed by the microcomputer and identified. If the part is misorientated a match does not occur. The camera may be moved slightly to see if it is the camera angle which is responsible. If this is not the case, then the system must order the robot's hand to move the object around to see if the proper orientation can be obtained. If the proper orientation cannot be obtained, the microcomputer decides that this object has no business being there and will instruct the gripper to separate it for human inspection.

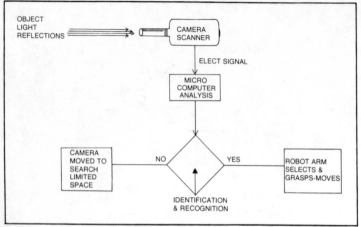

Fig. 9-7. A basic block diagram of a vision system for a robot.

Assume that a robot is using visual feedback to insert a guidepin in a hole. To provide position information, the system will track four features of the scene; the hole, the tip of the pin and both sides of the pin. These are three totally different features. A dark blob is the hole. A complex curved line is the pin bottom edge, and the sides are straight vertical lines. The computer knows there must be a certain specific relationship between the curve and the blob, and the straight lines and the blob, and these must occur simultaneously. It then steers the robot's hand to try to accomplish this.

We are now confronted with getting some dimensional coordinates of the object which the hand is to take hold of.

Notice that the Z coordinate in Fig. 9-8 is the distance from the camera to the object. The X coordinate is the left-right position, and the Y coordinate would be the forward-back position. Imagine an object coming along a conveyor belt. If the camera were exactly aligned, the object would move from $-X$ direction, through zero, and on out to the $+X$ direction. The Y coordinate might be zero all this time. The Z coordinate (the distance) will change as the object moves directly under the camera and then away from it.

Imagine the picture that might be seen by the camera. If we have a monitor screen to look at, we might see the dark shape appear at one side of the screen and vanish off the other

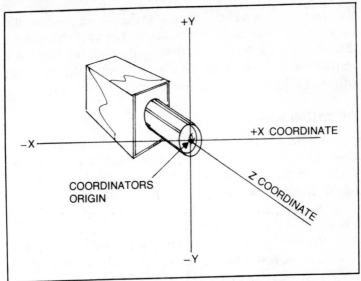

Fig. 9-8. The camera coordinates of a system for robotic vision.

end unless the gripper appears to grasp it. A pictorial representation of the situation is given in Fig. 9-9. The computer, which is directing the operation of the robotic arm, will have been informed where the objects are to appear, the shape they should have and of course their existence as denoted by a darker blob of light reflected from the object than from its conveyor. If the object is not in the position it should be in,

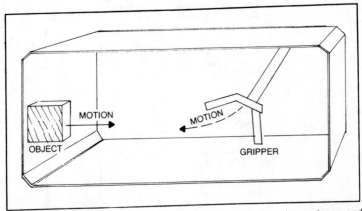

Fig. 9-9. An imaginary monitor screen shows the "object" and the gripper, and the motions of each.

359

then this might be a signal to put it in the proper position. If it does not have the shape the computer thinks it should have then the computer might stop the conveyor and have the robot re-position it. If this cannot be accomplished the computer will decide that the part is not useful and discard it.

THE VISION WINDOW

The computer cannot look at the whole screen all the time, so it finds it convenient to establish windows to get useful information concerning the object it is looking at. We might illustrate this as shown in Fig. 9-10.

The computer finds the hole blob where it should be, in window 2, and moving in a straight line toward window 5. It finds the bolt blob occupying window 1, window 4, and window 2. If the bolt were positioned properly, it could not be seen in window 4 at the same time that it is seen in window 1. So the computer has something to go on to adjust the bolt's orientation. If necessary the computer can shrink or expand the size of the windows.

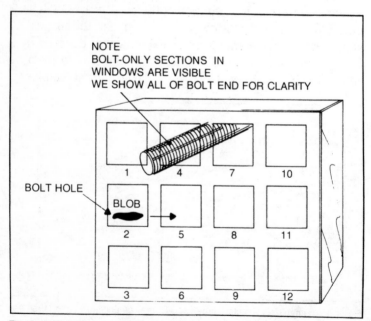

Fig. 9-10. Robot-Vision "windows" depicted on an imaginary display monitor.

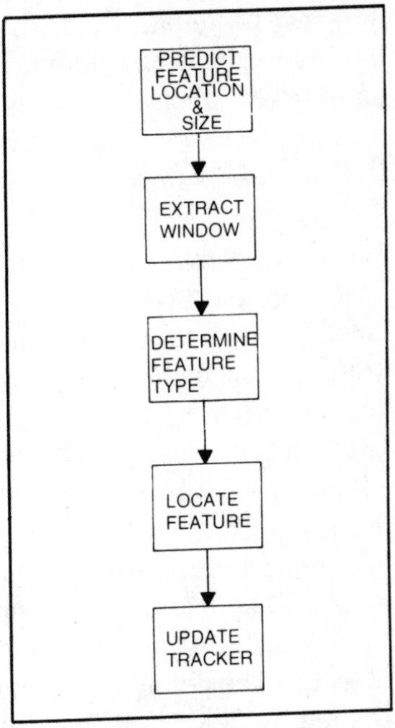

Fig. 9-11. A robot-vision tracking system block diagram.

Many robotic eyes are light-sensitive cells arranged in a matrix. Each cell produces a signal whose intensity or polarity is a function of the amount of light falling on it. The light-gathering unit is the lens system in front of the matrix.

Figure 9-11 shows what the seeing robot has to do in a situation where it is required to follow an object or track its position. The computer identifies a feature of the blob that it sees. This feature may be the round segment of the hole. It could be the straight parallel lines which tell the computer it is looking at the bolt. It could be the shaped curve of the bolt tip, which is a different curve from that of the hole. The system tracks the feature by causing movement of the robotic arm, hand, or camera.

Scanning a matrix window and producing a voltage according to the various shades of color from white, then digitizing this voltage and giving that information to the microcomputer is how the computer can get the information it needs from the robotic-eye system. The computer can then compare this digitized signal with memorized signals in its memory bank, and do something as a result of this comparison. Figure

9-12 may be helpful. Each window may be scanned separately in a high resolution system. The scanning process is just like standard TV technology.

USING THE MATRIX EYE

Scanning a light sensitive surface with a beam is one method of converting light into voltages. A better way is to have a grid which is composed of an incredible number of small wires running vertically and horizontally and then tapping into this matrix by means of a computer like system and reading the voltages produced at each cross-wire junction. General Electric has such an eye and we show it in Fig. 9-13.

General Electric calls their new camera the TN 2200 and it uses a charge injection device, (CID) to convert the light energy into an electrical signal. The system resolution can be enhanced by use of an appropriate system of lenses.

With the grid system the X and Y coordinates of the object will be defined by the electrical charge at the junctions in the grid field. The Z coordinate can be determined by the intensity of that charge.

SOME BASICS OF
ROBOTIC-VISION COORDINATE DETERMINATION

Figure 9-4 shows the military problem of trying to hit an airplane with an artillery shell. The first task in that problem is trying to define the position of the airplane in a three dimensional world. That is, its altitude, its azimuth, and its distance. Here we assume that the observer's position is the origin of the coordinate system. The airplane is flying along a path which is directly on the X axis, and it is approaching the observer at a constant altitude Z. As shown we could say the plane has at this moment an X coordinate of 18 units, a Z coordinate of 14 units and zero Y coordinate. The TV camera can certainly provide information on two dimensions of that space, but what about the third measurement. How can a TV camera give us range information?

The human eye is able to perceive distance because two eyes are used. There is some distance between them, and the brain can perform the necessary calculations (and we aren't

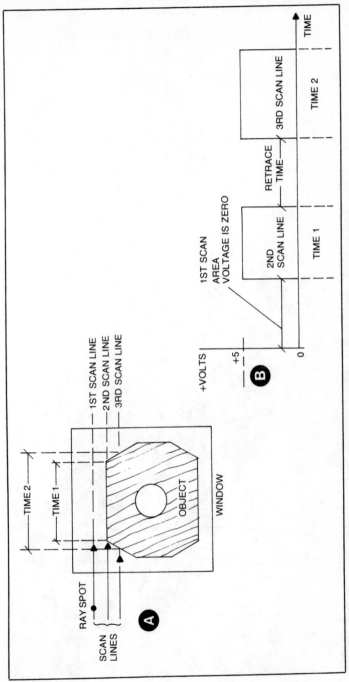

Fig. 9-12. The scanning voltages for three lines of a robot-vision scanning system. A hypothetical case.

Fig. 9-13. The General Electric Robotic-Eye. Notice the matrix at the end of the tube. It is very small but has a very large number of "cells" (courtesy G.E.).

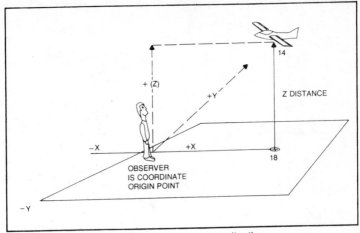

Fig. 9-14. A three-dimensional airplane position situation.

even aware it is doing so) to tell us the distance of an object. This system is called triangulation and, the answer, of course, yes, it can be used in robotic applications. Let us examine Figs. 9-15 and 9-16. Figure 9-15 uses a laser and one camera; Fig. 9-16 uses two TV cameras. They examine a given point or object through identical angles. The base distance between the cameras is known. A side and two angles are known. The other two sides can be determined. The range is then calculated. The angle is how far the head is turned. Polar coordinates are converted to Cartesian (X,Y) and then the elevation (Z) is found.

You will recall that we have to track the recognizable edge or curve of a particular shape. Our range finding system must be able to differentiate, in perhaps fractions of a millimeter, the distances to various sides of, say, a hexigon shaped nut. Whereas a camera might not be capable of this as Radar and Sonar are.

THE RECOGNIZER, THE COMPUTER, AND THE CONTROL SYSTEM

The recognizer is that part of the system which identifies some characteristic of the object. It may be the outline, reflectivity, position, or something of that nature. Figure 9-17 is a block diagram. If the recognition is satisfactory the com-

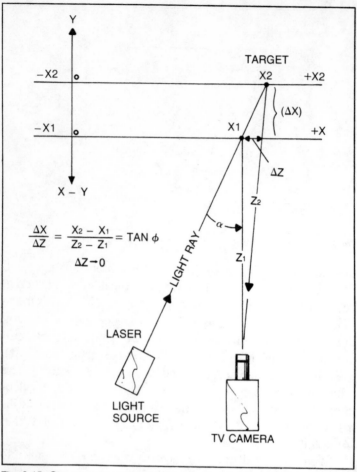

Fig. 9-15. One concept of distance measurement using a T V type camera imaging system.

puter orders the control system to do something to the object. If the recognition is not satisfactory then a new look is ordered. The system tries to identify the object as something which should be there, even though it doesn't have exactly the right position or orientation.

THE SYSTEM CONCEPT OF G.E. OPTOMATION

We have mentioned the G.E. CID video camera. Examine it in Fig. 9-18. It is not very large and it differs in configuration from the camera we have examined in Fig. 9-13, but the operation is the same. The basic concepts underlying the Optomation video system are three.

The first is visual observation with the CID camera. Because it has a very stable coordinate, system, it can effectively super-impose a stable and repeatable measurement grid over the object under observation.

The second is the recognition of the object by the extraction of some of its key features.

Finally, based on observation and recognition, some kind of decision is arrived at by the computer controller.

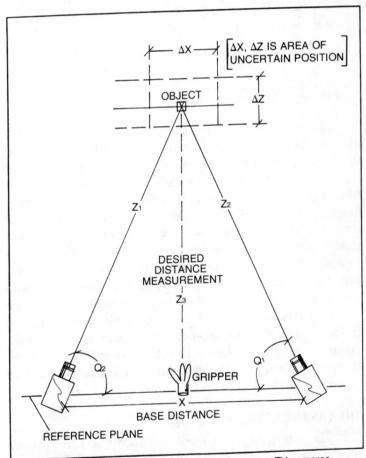

Fig. 9-16. Triangulation measurement of range using two TV cameras.

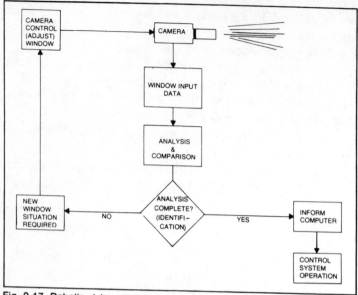

Fig. 9-17. Robotic-vision recognition system, block diagram.

In the system development a strategy has been employed which says that there will be two levels of viewing. In the Level I system operation the objects are placed in the camera's field of vision in a carefully pre-determined orientation. The system then obtains such information as length, height, area, and relative or absolute location. The relative location may be the position of the object with respect to the robotic gripper. The absolute location may be the object's location with respect to camera. In this system simple shape recognition can be accomplished by comparison of the viewed object with a feature set which has been memorized from a previous image. If desired a multitude of electronic windows can be used all at the same time, to make many measurements simultaneously. The Level II system operates on all the things the Level I does and also can examine randomly oriented objects.

THE CAMERA ROBOT INTERFACE

The interfacing unit feeds the robotic control system signals from the camera. The G.E. interfacing unit requires

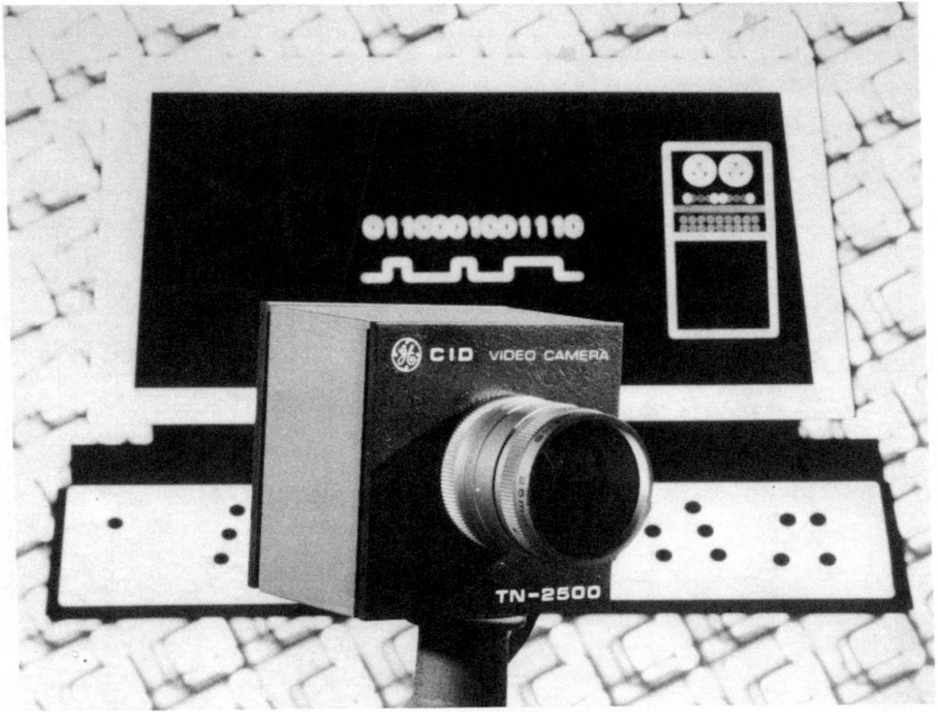

Fig. 9-18. The G.E. Optomation CID robotic camera.

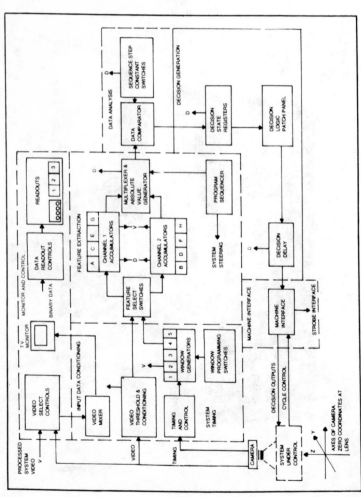

Fig. 9-19. A block diagram of the G.E. Optomation system (courtesy G.E.).

only that a cycle control pulse be sent to it. This pulse informs the interfacing unit that the camera data is ready to be accepted by the computing section. It also causes a new frame of video to be put into the feature extraction section. This, in turn, causes a new set of measurements to be taken.

There are 8 separate decision outputs available to the computer and control system from the interface unit. These may be in either one of two forms. They may be in TTL positive true logic levels or a one amp, 60 Hz control signal of either 110 or 220 volts, present when the output is true. A strobe interface is provided in this system to synchronize a stroboscopic lighting system with the input cycle.

DECISION GENERATION IN THE OPTOMATION SYSTEM

The contents of the decision state registers provide input data for the decision generation function. These registers are updated each time a measurement cycle is performed. Notice how the camera is used in this robotic application. It is providing both measurements of position and recognition of objects in its field of view.

Going to the decision registers we note that a variety of output decisions can be generated from the input decision states. A part may be rejected, or selected, on the basis of only height measurement. It can be rejected if the height is too large or too small. A part may be rejected because its width and total area are not within specifications.

A delay function is also available and output decisions may be delayed by as much as 255 measurement cycles. The delay interval is determined by an 8-bit binary number which can be set by the human programmer using another system of multipole switches on the programming panel.

In Fig. 9-20 we examine the system's control panel. At the top are the multipole switches and the top section is labeled "Control Panel". On the second broad row the first unit is the decision logic panel and next to it is a data readout panel. This panel shows the decision outputs, the binary values, the channel routing and a section to compare results and the decimal value of the binary readout. In the third row down, on the left, is shown the feature control section and the

371

Channel 1 and the Channel 2 blocks. On the right of this section is the sequence step constants section. On the bottom row we find the section for window control.

OPTOMATION SYSTEM OPERATION

The PN 2303 system has seven primary modules as you can see from an inspection of Fig. 9-19. These can be designated as; the camera unit, the input data conditioning unit, the feature extraction unit, the data analysis unit, the decision generation unit, the machine interface unit, and the monitor and control unit.

Fig. 9-20. The G.E. Optomation system control panel.

The camera unit is a logical starting point. It has a CID eye which means that it is composed of a mosaic of solid state light sensitive elements called pixels. The TN 2200 of Fig. 9-13 has 16,384 pixels, and the TN 2201 has 14,364 pixels. Pixel time is the time required to scan a single pixel or light sensitive point. Pixel time for the G.E. TN-2200 camera is 10 clock periods.

Frame time is the time required to scan a full array of CID elements and it includes the blanking time. The TN-2200 frame time equals 17,688 pixel times and the TN-2201 frame time is equal to 16,608 pixel times. The line time is that time required to scan a single line of pixels and it includes the blanking time. For the TN-2200 it is 134 pixel times and for the TN-2201 it is 48 pixel times.

As used in the schematic diagram the X,Y, and Z axes are the same as an oscilloscope's axes. X is left-right, Y is up-down, Z is bright-dim.

As we develop this subject we will be seeing some abbreviations, and so definitions of them will also be useful here. TTL is transistor-transistor-logic. LRR is line re-read. Inhibit Input is abbreviated II. EOL is end-of-line and refers to a CMOS output of the camera that goes negative when the last element in a line is selected. EOF stands for end-of-frame and this is similar to the EOL except that it refers to the last pixel of the frame. EOL and EOF are used to designate retrace. STROBE is flash lighting.

In the TN-2200 unit the pixels are located just 0.0018 inches apart from center to center. The camera will accept standard C mount lenses which have a fixed back mounting flange length of 0.69 inches. The camera uses plus and minus 15 volts dc at under 50 milliamperes, and there is a 12 conductor cable used to attach the camera to its ancillary equipment.

There is a relationship between a lens picture size and the actual size of an object. The camera under study at the moment has a chart relationship which defines the size of the object as seen by the camera in terms of a ratio of object size to distance. The focal length of the lens is the factor which governs this ratio, as follows:

Focal Length		Object Size/Object Distance
mm	inches	
4.5	0.18	1.3
6.5	0.26	0.89
9	0.35	0.65
12.5	0.50	0.46
16	0.63	0.37
17	0.67	0.34
25	1.0	0.23
35	1.4	0.17
50	2.0	0.12
75	3.0	0.08
100	4.0	0.06

To view a 6-inch object at a 4-foot distance, the object size/distance ratio is 0.5/4 = 0.125. Therefore, a 50 mm lens is appropriate for this application.

The video output signal of the camera can be converted to a TTL level signal with a threshold anywhere in the gray scale required. A simple conversion is shown in Fig. 9-21.

THE INDUSTRIAL VIDEO-INSPECTION SYSTEM

Figure 9-22 is a block diagram of the industrial video inspection system which shows us in a very simple arrangement how the robotic eye can be used gainfully. The signal processor shown is the computer. When it defines a part as

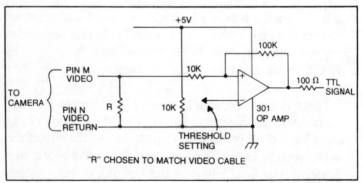

Fig. 9-21. A Video to TTL (transistor-transistor logic) signal conversion circuit.

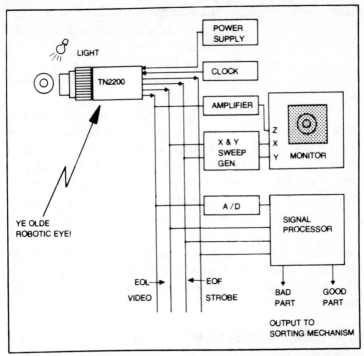

Fig. 9-22. An industrial inspection system using a G.E. robotic eye. The speed of scan is chosen to match the signal processor requirements, and is controlled by setting the clock frequency. The decision as to whether or not the part is *good* or *bad* is made by the signal processor which may be a mini-or micro-computer based system. An analog to digital conversion will be required for the video signal if it is to be processed digitally. In many systems a monitor view of the object is useful. A view may be presented by amplifying the video to modulate the Z axis of an X,Y monitor. The Z axis is the light intensity on the monitor tube presentation.

bad, the robotic arm and gripper are signaled to reach out, grab that part, and convey it to the re-cycling area. If the part is good, the robotic arm may pick it up and send it along on another belt.

THE CID CAMERA CONSTRUCTION AND OPERATION

The signal readout is accomplished by a half-address scheme which is similar in concept to the core memory readout. The row scanner applies an address signal to a given row, resulting in a half-address condition on all the row pads in that row. (See Fig. 9-23.) The column scanner then sequentially applies the appropriate address signals to each of the columns, one at a time. This establishes a half-address condition on all the column pads in each column in turn. When all the columns in a given row have been scanned the row scanner increments to the next row and the process is repeated.

The result is a video signal in a television-like format. This read-out is fast—usually 30 frames per second. Of course this output is analog in nature and so to use it in a microcomputer you have to feed it into an analog-to-digital converter. There are many types described in the current literature. If the signal is to be used directly by a robotic control system, the converter in Fig. 9-21 can be used to digitize it. The use of a video system for measurement of parts size and shape can produce really good accuracy. In the system described, it is claimed that an accuracy of up to 4 mils (0.004) inch can be obtained using microscope-optics systems.

FOLLOWING THE VIDEO SIGNAL

If we again refer to Fig. 9-19 we see that the video is immediately converted into a one-bit binary, black and white signal. The threshold level is adjusted electronically, but compensation for light variations can be made by adjusting the camera lens setting or the aperture opening. The balance of the equipment shown in the block diagram uses the binary signal as input except for the TV monitor, which uses the raw video. The system has a video mixer which permits the processed video to be superimposed on the normal image as an aid to setting up the system.

After the signals go into the feature and select switch where information from the window generators unit may be selected, the signals go the feature extraction section consisting of the accumulators. Let us examine the window feature in more detail.

The windows may be of any size or at any position within the field of view. They may overlap if this is desired. The position of each window is programmed by means of a set of 8 switches which set up binary numbers equivalent to the

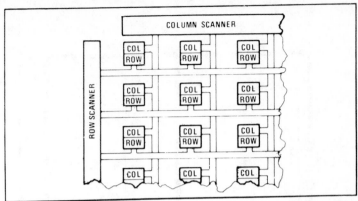

Fig. 9-23. The robotic camera eye's CID (charge injection device) row-column matrix system (courtesy G.E.).

column counts at the left and right edges of the window desired. They also set up the line counts at which the top and bottom of the window are to be located. The person looking at the monitor can superimpose a white level signal on the picture to show where the window is located. The output of the window generator is sent to the input of the feature extraction unit by means of a replaceable wire-interconnect header. Any of the five window generators can be assigned to any or all of the eight feature data accumulators, but only one window can be assigned to any one feature. As an example, window 1 may be assigned to both feature A and B, but windows 1 and 2 may not be assigned into feature A at the same time.

WHAT IS FEATURE EXTRACTION?

We can gain some idea of what feature extraction is by examination of Fig. 9-24. What the feature extraction function does is to count the black and white pixels within the boundaries established by the window generators. The specific data to be stored in the eight accumulators is determined by the user. The type information which needs to be stored is selected by setting the switches on the control panel. Each of the small square blocks contains eight single pole switches. In the block diagram of Fig. 9-20 we note that there are two accumulator channels which can be used to store information.

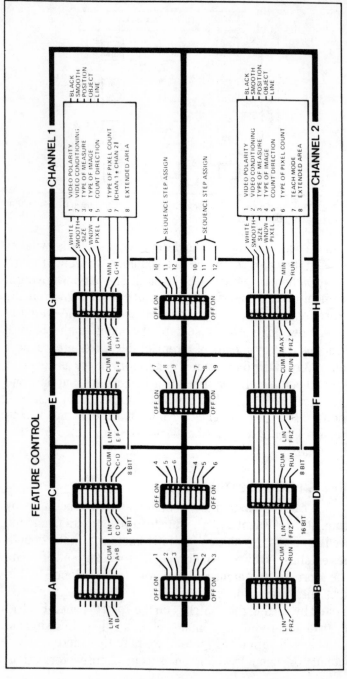

Fig. 9-24. The Feature Extraction Function control panel (courtesy G.E.).

378

The reason for this is so that you can arrange for the storage of the absolute values in two sections and then you can get sums and differences by comparing these values if this is desired or necessary.

The feature control unit with all of its switches does permit a great deal of flexibility. For example, you can obtain your choice of video polarity which can be important depending on the background of the object under inspection. You can also select either smoothed video data or raw video data.

To find the position of the object in the space viewed you can check either the position of the object in the window section or the edge of the object in the window.

To measure size the count direction will determine whether measurement will be made along the X axis or the Y axis. You can choose whether you want a linear count or a cumulative count. The difference is a linear count will give a width of the object along a single line within the window. If you choose a cumulative count, then you'll get the total area in the window. Total black and white pixels are in the two data accumulation units, one in each channel.

The contents of the feature data accumulators are time-multiplexed into the channel 1 and channel 2 output buses under the control of the 12 channel step program sequencer. This performs the total basic steering function of data for the entire system. Each feature data accumulator can be assigned as many as three program steps and any unassigned data effectively becomes zero.

THE DATA ANALYSIS SECTION

The data analysis function is accomplished by comparing the present data with the stored data. The contents of selected feature data accumulators are multiplexed by giving each a time slot into the output bus at each of the 12 program steps. An eight-bit bus conveys measurement values for each sequence step. These values are then sent to the data comparator in a sequence. At the same time another set of values which define the acceptance or rejection thresholds are sent from the twelve sequence step constants switches to the data comparator. The output signal is used to drive the mechanism which accepts or rejects the part.

More on Robotic Vision and Speech

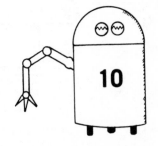

10

In the previous chapter we found that in the G.E. system the eye was a special camera which used a special matrix of light sensitive cells. In the system we are about to investigate, we find that a standard vidicon camera tube or an optional silicon-diode type tube is used. The system digitizes the video output of the camera so the information can be used by a microcomputer and its associated control system.

THE SPATIAL DATA SYSTEMS EYECOM II

This is a picture digitizer and display system. Its general operation and assembly are shown in Fig. 10-1. Notice in the diagram that the reflected light from the object goes through an optical system and impinges directly on the face of the vidicon camera tube. There, a beam of very small diameter scans the face of the tube and generates the video signal, which is analog in nature. The output goes to a display picture tube for a person's inspection and also goes to a digitizer and from there the information goes, via control, data, and timing buses, to the computer section. Most of the electronics are used to enhance, expand, or update the picture as seen on the monitor. The joystick can place a marker on the video monitor screen, and move it as desired. I call your attention to a block

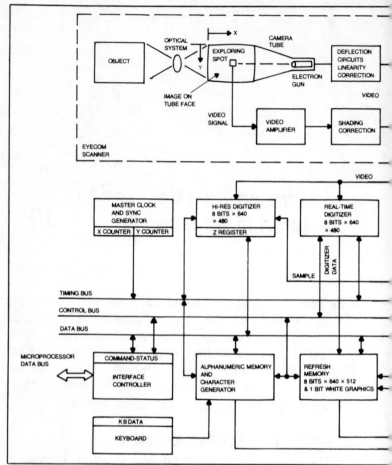

Fig. 10-1. A Block diagram of the Eyecom robotic-eye system (courtesy Spatial Data Systems, Inc.).

near the camera tube, the shading correction block. The circuits contained in this block provide the color enhancement and correction which can be accomplished in the EyeCom System. The system uses a scanning system which is compatible with standard television.

A light image of the object is produced on the photosensitive face of the camera tube by an optical system containing an ordinary camera lens. The system uses a standard C mounting which accepts television lenses, however, a universal screw type mounting for 35 mm camera lenses can be pro-

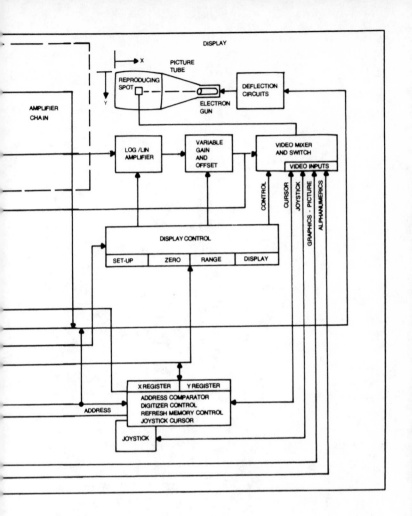

vided. In normal operation the image is stationary and the brightness at any point of the image is a function of the X and Y coordinate position on the tube face. The image brightness at each point is then defined as a third dimension or Z coordinate. This is converted into the video signal by repeatedly scanning the image with an exploring spot formed by the electron beam of the camera gun. Notice the EyeCom camera in Fig. 10-2. It may look familiar to you. It is relatively small and easy to mount in a number of either fixed or moving base positions.

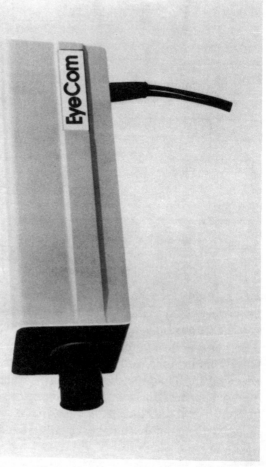

Fig. 10-2. The Eyecom-camera scanner (courtesy Spatial Data Systems).

THE INTERFACE CONTROLLER

All of the EyeCom functions are controlled through the control bus. This bidirectional bus transmits control signals to the displays, digitizers, and registers from the interface controller. It also sends status signals from the EyeCom internal units to the interface controller. Data is transmitted throughout the system on the data bus. The interface controller connects the microprocessor bus to the EyeCom bus. The keyboard used with the microprocessor connects independently to the interface controller. Figure 10-3 depicts the monitor screen unit and the microprocessor keyboard unit.

USING THE EYECOM SYSTEM
IN AN INSPECT-REJECT-ACCEPTANCE MODE

With proper lenses this system can inspect with tolerances as fine as 0.0001 inch! Using a Macro lens this unit is said to be able to check hole dimensions and locations in machined parts to a resolution of 0.0005 inch. The use of the EyeCom in this application is called the automated parts measurement application. Ultra small integrated circuits can be checked for accuracy, completeness, and tolerances in the PMS mode. One physical set-up for such inspections is shown in Fig. 10-4.

The smallest of circuits can be inspected for missing connections, for short circuits, for cracks in the base structure and so on. Since the data can be fed into the microprocessor, its computer can make a line by line comparison with referenced information gained from a perfect unit previously examined and stored.

In some visual inspection systems the table top upon which the part is mounted is caused to move by servo control commanded by the computer. The table top—and thus the part—can be moved.

The main requirement is good contrast between the table or mounting base and the part, and between the part mass and the hole. An automated operation is illustrated in Fig. 10-5.

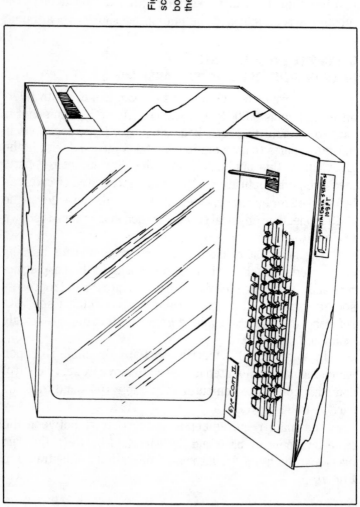

Fig. 10-3. Drawing of the Monitor screen and microprocessor keyboard of the Eyecom system. Notice the joystick on the right of keyboard.

Fig. 10-4. The Eyecom camera set up for inspection of Integrated Circuitry. An expanded picture is displayed on the monitor screen (courtesy Spatial Data Systems).

387

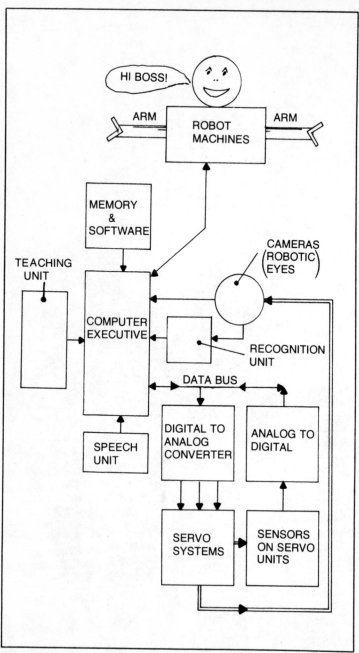

Fig. 10-5. A general concept of a robotic control system using microcomputers and robotic eye.

THE EYECOM DIGITIZER FLOW OPERATION

The X-register and the Y-register are loaded from the data bus with the location of the pixel to be sampled and digitized. The X and Y registers are continuously compared with the spot position as provided by the sync generator counters on the timing bus. When the scanning spot reaches the stored X, Y address the address comparator issued a sample pulse to the high resolution digitizer. The command to sample and digitize a pixel is received by the digitizer control from the interface controller through the control bus. The sampled video is digitized into an 8-bit binary Z-value and then stored in the Z-register where it can be accessed through the controller by the computer.

The physical electronics of the system are shown in Fig. 10-7. Notice that the circuit cards are so arranged that they are easily removed for inspection, replacement, or whatever.

Finally we examine a block diagram of the EyeCom color digitizer system in Fig. 10-8. If one wants to make a color display of something which isn't colored then the data from the picture display is converted to color by the optional color look-up table. Each of the 64 gray levels in the most significant 6 bits of the picture display memory is assigned a color. Colors are stored in the table by the digital computer. Colors are defined as proportions of the primary red, green, and blue values using 8-bits per primary. As pixel values are read from the refresh memory, they address the color look-up table. Color values are then converted to video signals.

In the block diagram we find four solid state random access memories in the refresh memory section. One of these is used for display of graphics and the other three are used for color pictures. Color pictures are produced by superimposing three separations in red, green, and blue on the color display picture tube. Each separation contains tonal variations represented by an array of pixels stored in the red, green, and blue refresh memories.

INVESTIGATIONS OF ROBOTIC EYES ATTACHED TO GRIPPERS

The idea of mounting a camera on the wrist of a gripper so it can see what the gripper is trying to grasp obtained results

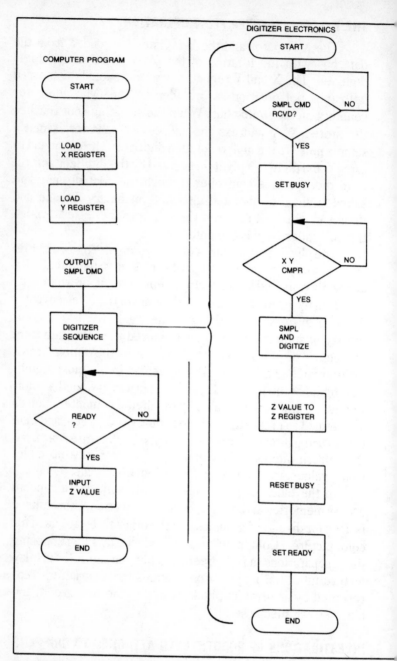

Fig. 10-6. Computer programming and digitizer flow diagram (courtesy Spatial Data Systems).

390

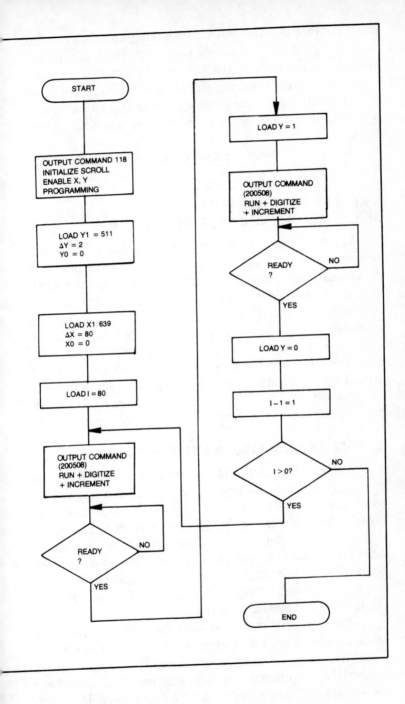

which indicate that this may be something which we can use to advantage if it is refined and improved. The future of this kind of system titilates the imagination. Smaller cameras, better lighting methods, ruggedization of the equipment, and good accurate algorithms might well bring about a whole new world of usefulness for the robotic arm and robotic eye systems.

More on this research is in the 1979 winter issue of *Robotics Today*.

A NOTE ON ONE INDUSTRIAL DEVELOPMENT TREND USING ROBOTIC VISION

In Japan, where there is great effort being made toward development of the automated factory, the development of the two-handed robot using three dimensional vision has been progressing for some years. The use of such robots able to coordinate the use of two hands and multiple fingers in the assembly of batch products is under intense study. At the University of Waseda, Tokyo, a robotic machine has been built which has two hands, two legs and two video camera eyes. This robot can recognize some speech and can speak to a limited extent. The use of two video cameras give a three dimensional view of objects.

IS THERE AN ADVANTAGE TO A TALKING ROBOT?

We know it is technically feasible for an electronic circuit to recognize and accept human speech and to generate some kind of response to those sounds. We also know it is possible for an electronic circuit to generate human speech so that it conveys intelligent messages back to us. These circuits are available at the present from Telesensors, Radio Shack, Texas Instruments, and other suppliers. We have already discussed talking units, so we will focus on voice recognition and human acceptance.

Obviously, for a central computer doing batch work voice communication with each terminal or user is not efficient. A high-level language is better. This is also true of industrial applications where more than one robot is used and the actions of all concerned need to be coordinated. However,

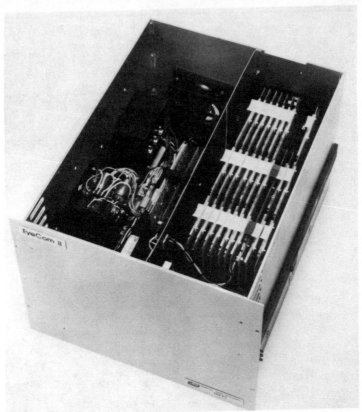

Fig. 10-7. The Eyecom electronics package with removable circuit cards (courtesy Spatial Data Systems).

individual and domestic robots might become much more efficient with voice recognition. The goal is to give the user the maximum support rather than to try to optimize the utilization of the computational resources.

English is not an exact language. Some people say "cut off the light," some say "kill the light," and some say "turn off the light." If we assign special words for each task, we are actually inventing a high-level language, and have gained nothing by adding voice recognition. In an effort to compromise Ruly English was developed.

Ruly English uses regular English syntax but limits each word to one meaning. It is not a colorful language, but is precise, and immediately understandable. One might say it is

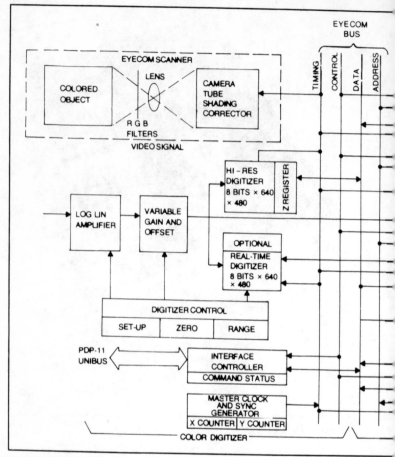

Fig. 10-8. The Eyecom block diagram of the color system (courtesy Spatial Data Systems).

a subset of the English language. It is possible to pictorialize these ideas in an illustration. See Fig. 10-9.

SOME GENERAL RULES FOR
VERBALLY COMMUNICATING WITH ROBOTS

Sanyo Electric has demonstrated a television receiver that responds to voice commands to turn itself OFF and ON and to switch stations. The receiver intelligence unit does this by comparing the voice input from selected persons to voice patterns which have been stored in its memory. At the

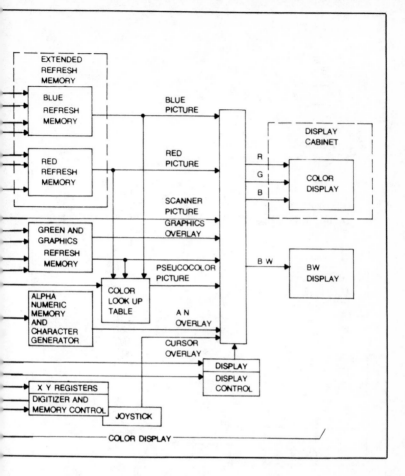

time of its demonstration this unit had a 30 word capability in vocabulary and could distinguish the voices of two different people. An expansion on this concept would permit people to play computer games using vocal responses, according to the research information which is available. However, this unit has problems identifying a "learned" voice when it has a cold. This could create problems in the workplace.

There are other problems. Many humans wouldn't mind working next to a machine they can't program, but few will work next to one they can't stop in an emergency. So the robot would have to accept only certain voices for programming, but any voice for operation. Of course, by the time these

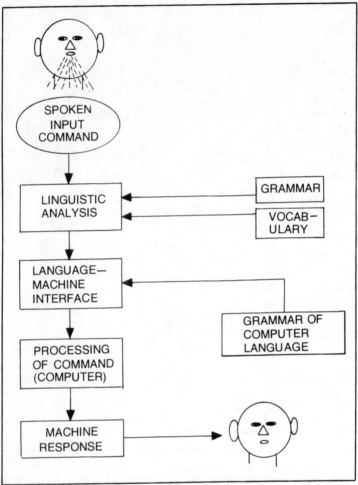

Fig. 10-9. A block diagram of a natural language man-machine communications system.

robots are common in factories, today's children will be the workers (maybe) and growing up with talking TVs and learning aids may cause them to accept all this much easier.

The popular adventure games are an example of trying to get a computer to understand English. The earliest ones used two word commands. This was not a noun and verb as you might expect. Since these were commands, the noun was the implied "you." To command was a verb and either a verb modifier (GO NORTH) or a direct object (THROW KNIFE). If the action implied an indirect object, as in this example, the computer asked "TO OR AT WHAT?"

The computer looks to see that it "knows" the verb and object. If not, it tells you it doesn't understand. Next it checks to see the verb is applicable to the object. If not, it tells you so! (CAN'T EAT HOUSE!) Finally it checks to see if you have the object. You can't throw a knife you don't have! Since the first simple programs, routines have been developed allowing the computer to pick the important two words, plus indirect object, from a complete English sentence.

Fig. 10-10. Texas Instruments Speak and Math machine (courtesy Texas Instruments).

There are four primary significant models of language and these are; the fixed sentence model, the *finite states model,* the *syntactical model,* and the transformational model. These are described in detail in *Syntactic Structures* by Chomsky, (Monton & Co, the Hague, Paris, 1957). The first two are the most applicable to present voice recognition technology.

In the fixed sentence model there is simply a list of all the allowed sentences. One defines each sentence with a mathematical term. If we develop a whole list of sentences which are unique then when the computer receives an order it essentially compares that order to that series of sentences and chooses one which matches or most nearly matches the command.

The finite states model is essentially what the adventures use.

SOME CONSIDERATIONS OF AUTOMATIC SPEECH RECOGNITION BY MACHINES

Once the meaning of a sentence has been decoded, there is still the problem of identifying the speaker.

There are two general ideas for recognition of human speech. These are a system where the pitch and spectral characteristics can be analyzed and related to some person, and a system where the electronic circuitry can extract phonemes and from a tabulation of these phonemes identify the different persons speaking.

In a study at the Rome Air Development Center, Griffiss Air Force Base, New York, an analysis of human speech was conducted. It was found that speech sounds are primarily of two types—voiced and unvoiced. For voiced sounds such as the vowels the vocal chords vibrate and produce a line spectrum which has a fundamental frequency in the 70-180 Hz region for males, with harmonics extending to 8,000 Hz and even higher. The unvoiced sounds are produced by the turbulence of the air stream as it exits the mouth and not from the vocal chords at all! There are also sounds such as Z which require both techniques.

There are about 40 basic phonemes in the English language. There have been detailed studies to find out just what identification characteristics there are in human speech, so we can make machines do what we do, recognize who is speaking.

A block diagram which shows a system for recognition of human speech, based on pitch and spectral characteristics is shown in Fig. 10-11. In a typical operation using this type system for speech recognition, the speech is fed into the spectrum analyzer and the pitch extractor as shown. The output of these units is sampled every 20 milliseconds, digitized, and sent to the computer. The computer knows what characteristics it is looking for because it has previously stored patterns of speech.

The system shown in Fig. 10-11 uses some clues to identify whom is speaking, or has spoken. One set of clues is to use the speakers fundamental frequency—the average pitch and the maximum and minimum of his speech tonal range. A second set of clues may be extracted from the speech spectrum, the measured value of the highest peaks of sound, the ratio of the amplitude of the sound, to the valley amplitude following the highest peak and the measurement of the peak amplitudes, and where they occur in the spectrum. One manner in which the information may be used is to form multi-dimensional histograms. Then during the recognition process the summed-squared-deviation between what is in the computer memory and the sound now coming in through the system can be compared. If the sound being evaluated has the same or close to the minimum summed-squared-deviation of the memorized pattern, then the computer decides that this is the speaker whose histograms are being used as a reference. Refer to Fig. 10-12.

Examine a block diagram of the method of speech identification that uses the automatic phoneme recognition concept. This can be illustrated as shown in Fig. 10-13. This block diagram is the one which would use the algorithm illustrated in Fig. 10-12. Basically, this equipment would make use of analog threshold-logic operating on the rectified

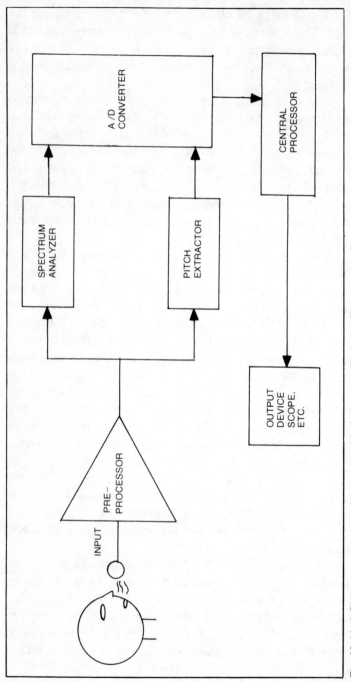

Fig. 10-11. A Pitch-spectrum recognition system block diagram (Rome Air-Development Center).

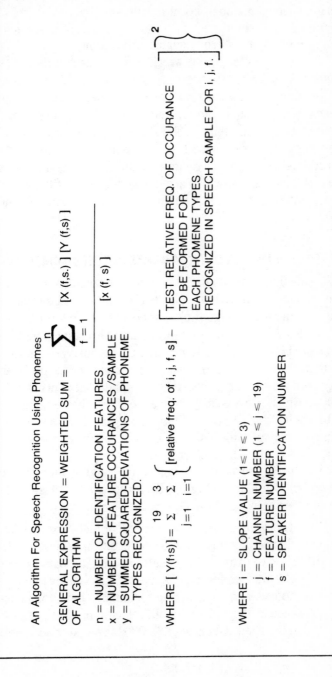

Fig. 10-12. A phoneme speech-recognition algorithm. An algorithm (equation) which uses the weighted sum of the summed, squared-deviations.

output of the filter banks to provide such features as spectral energy, spectrum slopes, local maxima-minima, transitions, sequences, and simultaneous occurances. All of these features are used in speech recognition systems.

In the RCA equipment which is illustrated in the block diagram, Fig. 10-13, the input is conversational speech and the equipment is used to recognize, and segment from the speech, a set of phoneme-like elements. During the occurrence of each such element the slope of the logarithmatical spectrum was employed to measure the speaker's characteristics-of-speech by integrating the spectrum slopes of the analog signals, which are derived from the filters, and then quantizing these to three levels; positive, negative, and zero.

TEXAS INSTRUMENTS DEVELOPMENTS IN SPEECH CIRCUITRY

Texas Instruments has done much advanced research in both speech synthesizing, speech recognition. They hold a patent for a single stage digital speech-synthesis filter. We have indicated how important filters are in breaking down the components of speech so that they may be analyzed. In fact, the unit shown earlier represented a major price break in voice synthesis using a breakthrough called linear predictive coding, which we discussed in an earlier chapter. Figure 10-14 shows the circuitry. Since using the inverse of this technique may provide a similar breakthrough in speech recognition, we will take a closer look.

We have indicated the algorithm solving unit in the microprocessor block of Fig. 10-15. It represents the summation of a series of terms from $i = 1$ to $i = n$.

"Brute force storage of speech signals can be accomplished by sampling and converting speech at an 8 to 10 kHz clock rate. This results in a digital data rate of 100,000 bits per second of speech. Pulse-coded modulation codecs and companding techniques have found acceptance in all new, all digital, telecommunications systems, but their data rate of 64,000 bit per second is still high.

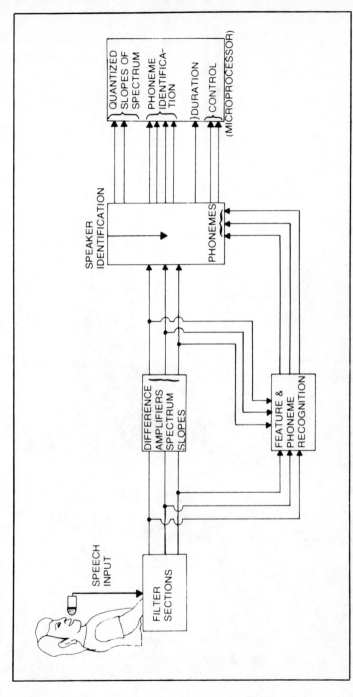

Fig. 10-13. A possible robotic voice tract, configuration. This actually is from TI's Speak and Spell device (courtesy TI).

403

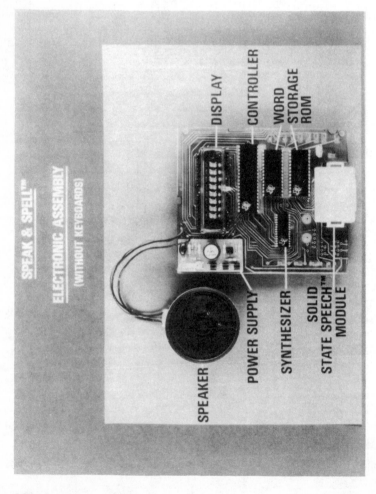

Fig. 10-14. How it could be done, speech recognition by phonemes (courtesy Texas Instruments).

404

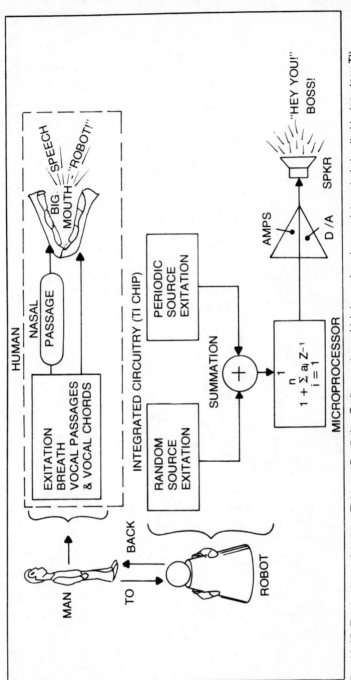

Fig. 10-15. The basic model of the TI Linear Predictive Coding system which is produced on an integrated circuit chip (courtesy TI).

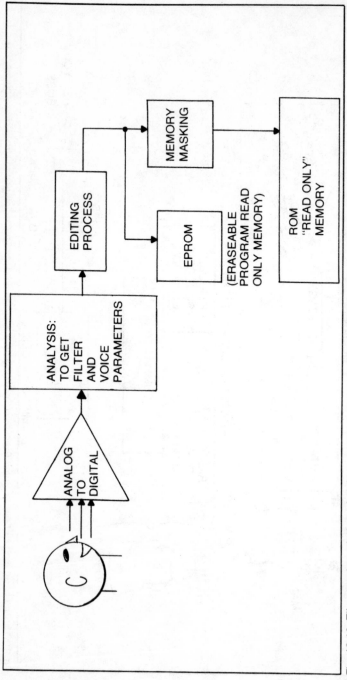

Fig. 10-16. TI's speech analysis and coding process to get proper speech formants into memories (courtesy TI).

"Linear Predictive Coding produces speech quality nearly comparable to either of these techniques, yet it only requires about 1,200 bits per second. With the advent of 128K bit and larger storage devices, LPC packs minutes of high quality speech into memories that could hold only one or two seconds of speech using other techniques."

The secret is that LPC predicts the parameters of the next speech sample from a linear combination of the values of the preceeding speech samples. This is possible because of redundant information in speech signals caused by a limited number of the formants in the human speech production system. LPC essentially eliminates unnecessary information in human speech and keeps only the data required to drive the synthetic model.

Speech sounds are recorded and digitized using special recording facilities that virtually eliminate background noise and preserve a high signal-to-noise ratio for the digitizer. Then the sophisticated signal algorithms of the type we have examined previously, are used to extract pitch and energy and slope information from the recorded data. Using a minimum-mean-square predicted-error criterion, the computer

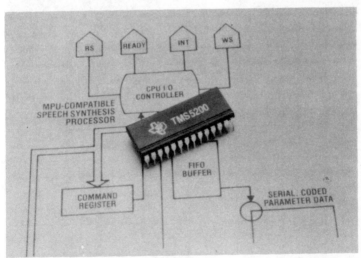

Fig. 10-17. The TMS5200 speech chip from Texas Instruments. FIFO stands for first-in-first-out and relates to signals into the buffer.

program computes an optimal set of coefficients for the vocal tract filter of the model electronics unit. The response of this filter, then, when suitably excited, will closely resemble the original speech model

In the next stage of the LPC system a frame repeat analysis takes place. Here, similar sets of reflection coefficients (40 bits each) are eliminated from the data stream and replaced with one-bit "repeat" code signals. Stop codes are added at word-phrase boundaries, allowing the synthesizer to "speak" a complete word or phrase and stop automatically with no supervision from a host microprocessor.

After thousands of samples of speech are analyzed by expert speech persons, those samples judged to be suitable, from a pronunciation and diction standpoint are ready to be committed into a "memory image" format. The fully analyzed data-set is encoded according to the pitch, energy, and filter parameter coding tables of the selected synthesizer. Then the final speech product is ready for listener tests and acceptance or rejection for various consumer uses. Figure 10-16 is a block diagram of the speech analysis and coding electronics

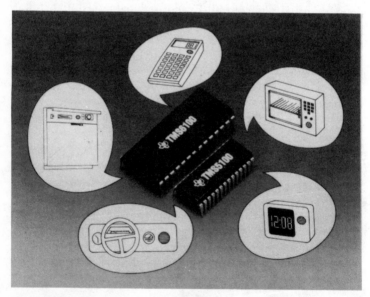

Fig. 10-18. Some possible uses, in addition to robotics, for the TI speech chips (courtesy TI).

system. This makes robotic speech a reality. It uses very little power and requires very little space. Figure 10-17 shows such a speech chip from TI. Of course, it has other uses too, as shown in Fig. 10-18.

Let us hope that TI continues to experiment with speech so that someday a single chip voice recognizer will be available. Because we like to give verbal commands,eventually we will have robotic units which respond to this type of input. It would be nice if they were cheap.

Ranging Systems  11

Polaroid's auto focusing camera uses a sonic ranging system. The unit has the ability to range accurately over reasonable distances is rather immune to extraneous ultra sonic sounds, and does not emit an audible sound. Polaroid offers a kit (Fig. 11-1) which can be obtained for experimentation. Write to: Ultrasonic Ranging Marketing, Cambridge, MA 02139.

PRINCIPLES OF CIRCUIT
OPERATION POLAROID ULTRASONIC RANGE SYSTEM

Sound waves which are compressions and rarefactions of the atmosphere, travel like the waves of the ocean at around 343 meters/second at 20 degrees centigrade. This converts to 1125 feet per second, which is the figure most of us are more familiar with. The speed of sound varies with humidity and temperature but is relatively free from effects of altitude.

The signal fall off is the inverse square of the distance traveled $\frac{1}{R_2}$ where r is the range or distance. And when there is a reflected echo, as when they are used in a distance measuring mode or ranging application then the echo also reduces in intensity. Thus we get $\frac{1}{R_4}$ as the total fall off ratio.

We know that since the energy travels at a specific speed, we can measure the time it takes for it to go to a given target and return. We can then multiply the time and the speed of travel and we come up with the round trip distance. Divide this by two and we get the distance to the object.

Emission of a single tone may permit extraneous notes at that pitch to enter the system. But a "chirp" signal, the emission of several tones, can override this difficulty to some extent, also "chirp" gives the detector of the system an ability to look at several echoes at different frequencies, or just one which may come from a secondary frequency in the "chirp" pattern. If just one tone were used, there might not be an echo, as the reflecting material might absorb the signal totally! Polaroid uses a four tone "chirp" at frequencies of 60, 57, 53 and 49.7 kHz to overcome this kind of difficulty. A very high reliability of echo response has been obtained in their system.

A pamphlet has been prepared by the Audio Engineering Society, 60 East 42nd Street, New York, NY 10017, titled *Polaroid Ultrasonic Ranging System,* and this can provide you with more details of the circuit operation, and some other considerations which were mandatory when developing this system.

USING A SOUND RANGING SYSTEM IN ROBOTIC APPLICATIONS

It is not necessary for a robot to be "intelligent" to make use of sonar anymore than it is necessary for a bat to be intelligent. Both are essentially blind, and the sonar reflection triggers reflexive action.

Such a sonic system can be used to trigger relays which have been arranged to control the drive motors, the steering motors, headlights, voice track, or voice synthesizer. An industrial application might be sonic ranging in a gripper. Like the photo-electric eye system, it can provide information that objects are on a conveyor belt, their spacing, and even the distance they are placed from the edge of the belt. Advance information of this type can be used to cause the microprocessor to adjust the robotic arm position to meet the on-coming objects precisely. And, unlike the photo-electric type device, the ultrasonic system needs no receiver other than its own transducer, and gives ranging information.

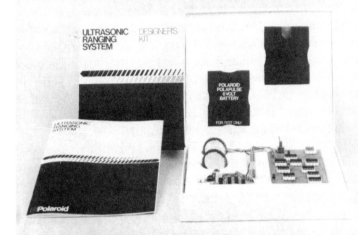

Fig. 11-1. The polaroid ultrasonic ranging kit includes two instrument-grade Polaroid transducers, a modified Polaroid ultrasonic circuit board, two Polaroid Polapulse 6-volt batteries, a battery holder, a wiring assembly and a technical manual (courtesy Polaroid).

But there may be some disadvantages. If the robot is working in an environment where there is a high sound level, and a multipitch sound spectrum, then it might get signals which would cause malfunctioning and problems. It is possible to filter the sonic signals (as Polaroid does) so that most other tone signals or ultrasonic signals are not received in sufficient strength to cause alarm, but, in a machine shop, for example, where tooling, grinding, etc., are taking place, you will find spikes of ultrasonic energy all over the low frequency spectrum to 100 kHz or even into the low rf frequency range. Look at the block diagram of the experimental demonstration circuit board package of the Polaroid Ultrasonic ranging system in Fig. 11-2. The range of the Polaroid system is from 0.9 foot to 35 feet, and that ranging information is presented on a digital display. Of course one could intercept the signals to the display and use them to activate other devices if this is desired.

A hobby robot will, no doubt, use TV camera type eyes, a

voice synthesizer for speech, and a multitude of sensor devices to provide information as to where it is, what it is encountering, and what is around it. Some persons, talked to in the preparation of this book, said that they find, in the larger toy stores, a multitude of electrically driven cars, trucks, tanks, and such, which make ideal starting units for robot development. Some, of course, start "from scratch" and built up from just an idea. Such was the origin of GARCAN which was illustrated in an earlier chapter.

John Cledhill discussed his robot in *Robotics Newsletter,* Sept. 1979. Some of what he said is interesting to us at this time. "It's surprising that most of the complexity of a computer (in robotics) is a result of the necessity of communications with humans. A dedicated, on-board robot computer doesn't have to have a keyboard, address/data light, etc., if you have a separate system for generating software. In my robot, decisions are made by a SC/MP-II microprocessor from National Semiconductor. But one could use the Motorola 6802.

My robot's basic senses begin with a bumper that encircles the frame and wheels. Eight microswitches provide input identifying if and where an object has been struck. The 8 microswitches for one 8-bit word are ANDed with constants by the computing section, so it knows which switches have been closed. The computer then knows what corrective action is necessary.

Since number 2 will be sonar using a National Semiconductor chip LM1812. The computer will tell the chip when to transmit and then time the echo response. Extreme accuracy is not necessary, as this will be used only to find out if there is anything in the field of view. Once something has been found in the field of "view", the bumper becomes the back-up device and will generate the operational signals causing whatever action is necessary.

This analysis and excerpt from that hobbyist's report gives us some indication of one way ultrasonic systems might be used in a very secondary role. With the Polaroid unit and with proper circuitry, there is no reason why ultrasonics can't be used as the primary control sensor. The Ultrasonic ranging

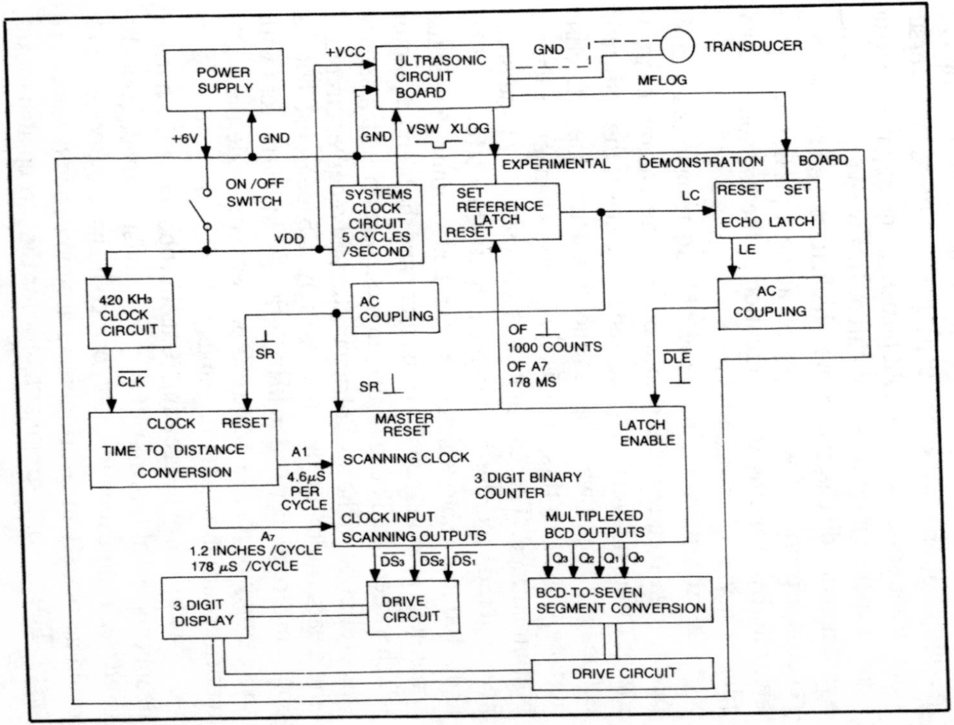

Fig. 11-2. The block diagram of the Ultrasonic ranging system of Polaroid's demonstration board (courtesy Polaroid).

system of Polaroid might be useful. With Ultrasonics your robot would not have to strike anything to get a signal that it was approaching something, or getting close enough to stop moving or to turn away!

MESOTECH SYSTEMS LTD of Vancouver, B.C. makes a mode-952 Bottom-Scan-Profiling-Sonar with some features that might be useful to think about in connection with robotics applications. Of course the transmission of sound in water is somewhat different than the transmission of sound through the air, especially with regard to the frequencies used. We know that some frequencies propagate very well through water and others do not. We have no correlation for the propagation of these sound frequencies through air versus other frequencies which might be used. One might have to do some experimentation in this area. In any event, the sonar system under discussion is one which, from a single location, records a profile of water depths along a particular line of bearing. By changing the orientation of the transducer, profiles of several lines can be made.

This transmitter unit, the sonar head, transmits a narrow beam, high frequency acoustic pulse. The narrow beam is swept through a vertical plane, and, of course, the returning echoes are timed which gives the distance to the reflecting surface, which, in this case, is the ocean, channel, or harbor bottom surface. The complete system consists of the sonar head and the recorder case (which charts the bottom profile) and a 100-foot cable which connects the two units. Inside the sonar head are the acoustic transducer and the stepping motor which sweeps it. There is also an inclinometer which senses ships rolling, and the sonar transmitter and receiver circuitry. The beam width in this unit is only 1.5 degrees, but its operating frequency is 360 kHz. It ranges to 160 meters with an accuracy of plus or minus .5 percent. Its angle measurement resolution is plus or minus 0.75 degree.

What is very interesting to us in robotics is the narrow beam concept of this unit. If such an acoustic device could be used in air, on a mobile robot, one could have a device which could accurately measure objects around it, and distances to them. The use of a stepping motor for the sweep also means that a microprocessor memory unit would be able to determine exactly in what direction the transducer head was pointing. In essence, we might obtain good or better accuracy than with a radar unit.

In some applications the Mesotech type unit may be used to keep track of underwater robots that are free to move in their environment, yet are under observation and control of a parent ship on the surface above. No cables are necessary to send control signals. Coded acoustic pulses can convey this information, as well as give range and bearing from the mother ship's reference position. The TV signals, however, do at present, require cables. It could be a problem knowing where your underwater robot was, and also communicating with it, unless you have very good sonar systems in use. There are some frequencies which tend to "channel" in water and go a long way with a reasonable amount of power. "Windows" these frequencies are called. It is safe to assume that 360 kHz is such a sonic window.

A RANGING SYSTEM USING LIGHT

One ranging system which uses the reflected light from an object to adjust a camera lens into focus has been incorporated into the Japanese Sanyo Autofocus Camera Control System, in their ES-44XL-VAF super 8 camera. In this system, basically, two mirrors are used to see the object through the camera window. The light from the object is reflected through a prism onto a photo-electric detector element which, in turn, sends the voltage generated from its two elements into a circuit which measures the balance. When the balance is exact a signal is generated which is then compared to a signal which indicates the position of focus of the lens. If the two signals are not in agreement, a small motor is energized which then adjusts the focus. Figure 11-3 shows a skeletonized view of the system.

The use of a prism to focus the light rays onto a silicon type photodiode (actually two of them) is easily seen in the illustration. The use of the prism to get two beams of reflected light can be better understood in the inset of the figure. I titled this chapter Ranging Systems. This system

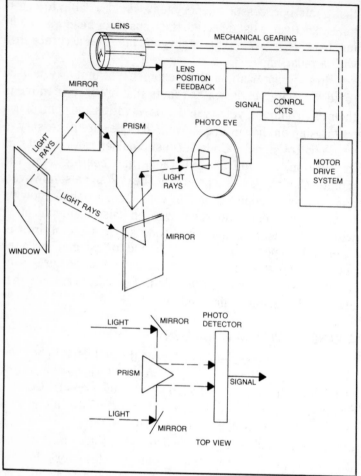

Fig. 11-3. A sketch view of Sanyo's autofocus system for cameras.

would calculate a range to an object by use of the focus information.

There would be problems adapting this system into a robot for its eyes. You should have a good illumination level, and that is not always possible. Of course you might incorporate an automatic lighting system into the robot so that a photo-illumination bright light will come on when the robot starts to move about. Then, it could get a good reflection, even if this causes some temporary blindness in the people it confronts!

Numerically Controlled Robotic Machines

12

It is essential that we examine what an NC machine is, learn how it is operated, and how it fits into the general picture of robotic machines. It is a machine which is controlled "by the numbers". What that means is that it is a lathe, a milling machine, a drilling machine, or something of the sort, that is programmed to move the cutting tool, drill, or whatever, exact distances to within .0001 inch to perform whatever task has been programmed for it, using a perforated tape containing the movement commands and instructions. A computer, naturally, is located in the system, and it receives the instruction numbers from the tape and knows what they mean.

For example, if a drilling tool is to move one inch on the X axis, over some plate which is to be drilled, and move two inches up the positive Y axis to exactly locate the center of the hole to be drilled, then the tape instructions would consist of two numbers. If the drill is to go down one inch to pierce through the metal plate at the specified position, then a third number is added to the instruction sequence. Since these instruction numbers are always in the same sequence, that is, the X axis instruction, the Y axis instruction and the Z axis instruction, then all that is needed is some way to space these commands on the perforated tape, which is read by an optical reader. This gives the commands to the computer which, in turn, controls the stepping motors, or the closed loop servomechanism, to make the machine elements do what the instructions say it should do.

The way the commands are spaced is by using a tab key on the tape perforating keyboard. This, like a typewriter tab key will cause the instructions to line up properly by forming a special kind of "end of command block" code which lets the computer know that another set of movements will be the next block's numerical grouping. After the machine drills the hole, as previously specified, it moves to a home or starting location where the coordinates are (0), (0), (0). The next command, which might consist of the numbers (5000), (6000), and (1000), tells it what to do next. See if you can interpret what the above figures mean using the information we have already given you. The tape used to store and present the command-instructions to the computer may look like that shown in Fig. 13-1. The numbers are in Binary form, i.e., the rightmost column is the 2^0, the next, 2^1, the next 2^2 and so on. Thus the position of the holes in the tape specifies a given number, and that number is a command for the machine tool movement. The rows of holes come in a specific sequence, as previously stated, which tells the machine's computer control brain how to move the tool, first in the X coordinate, then in the Y coordinate and in the Z direction. This, as you can realize, gives the position of a point in three dimensional space.

CHANGING TOOLS AND RANGE OF OPERATIONS

The NC machine can do marvelous things. If a source of tools is supplied to it, on a circular retaining unit, for example, it can, when so instructed, change its operating tool for each required operation. What it does is to move toward the circular retainer, which, itself has rotated under computerized command to a given position, attach its working head to the tool in a specific position, then the head is moved back to the work and to the position where that tool is to be used. When the task for tool number one is finished, the head will return the tool to the circular retainer, which grips it, the retainer

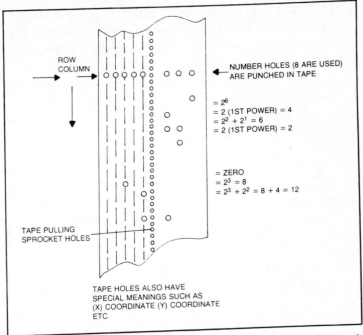

Fig. 12-1. A sketch representation of an NC machine's control tape. The tape is punched with holes which are converted into a binary number across the rows. Many rows of holes produce many numbers which designate the machine's operation. The column in which the punched hole appears governs its numerical or "special" operational meaning.

rotates again bringing a new tool into the acquisition position, and the head latches onto this new tool, and repeats the working procedure. A very wide range of operations from milling, to drilling, to smoothing, to contour shaping, can thus result.

FUTURE OF NC MACHINES AND ROBOTICS

Some of the scientists who develop and design these machines believe that in the future there will be a marriage of the NC machine and the robotic machine. The NC machine will be tended by an industrial robot at each advanced work station. All control data will be verified by computerized graphics. Sony has a color TV plant in Taiwan that produces a package product, ready to ship, with less than $1.50 invested in labor. Much of the work is done by work stations like this.

A reference on NC machines which might be of interest to you is *NUMERICAL CONTROL PART PROGRAMMING* written by James J. Childs and published by Industrial Press Inc.

Ideas and Discussions

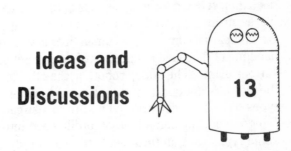

13

This last chapter will be a series of small paragraphs, each presenting an idea dealing with the design or concept of building a robot. Some are cheaper ways of doing things we have already discussed, others bring up factors that need to be considered in a society populated with robots. The first story will give you the flavor of the chapter.

LAWS

In Japan, an attempt to bomb the home of a syndicate leader with a radio controlled helicopter, toy variety, was foiled by the police. We may have to have some controlling laws governing the functions and operations of hobby robots as they increase in capability.

TYPES OF HOBBY ROBOTS

In the general case a robot is constructed to prove that one has the intelligence and mechanical-electronic-electrical-physical capability to make a machine that resembles in some way, a person, or the actions of a human. Thus, if nothing else, the construction of a robot of the hobby variety is a challenge, and makes one investigate most of the fields of science and engineering before the task is accomplished. We

423

might say, then, that we build hobby robots to learn, and to fascinate ourselves and our friends.

Many persons and companies as well as scientific institutions are making, or developing kits for, or are investigating the use of, hobby type robots. This type robot is distinguished from its industrial brothers in that the industrial robot has a definite task to perform, or many of them, and it does this well and for long periods of time. Tasks as demanding as making printed circuits and inserting parts on such circuit boards, something a human tires of very quickly, to the handling of thousands of pounds of equipment or materials in various plants. The hobby robot, on the other hand, is just for the pleasure or education, or satisfaction of the creator. Some types have been constructed just to investigate how far one can go in putting some kind of intelligence into a machine of some type. That, in turn, leads to much study and research. But for the many units which simply run around, or draw lines on something under computerized command it seems the satisfaction is simply to have built such a machine. It is our belief that with technology at the current state of the art that it is, even hobby robots should do something useful.

One robot which has been built and named Midnight Special, uses various types of light sensors to get information about paths through mazes. The information he derives is sent to a computer where it is stored and analyzed. When the Midnight Special has made one trip through such a maze it then can repeat its path without any deviations or problems or trial and error efforts. This kind of operation is said to be one in which a robot has been given some kind of intelligence. Certainly, it is a system in which the robot learns by its mistakes. Hobby robots should have this kind of capability.

WHO?

The robot builder comes from a diversified segment of our population. From the college professor to the high school science major. We found it interesting that a former college Physics professor found pleasure constructing hobby type robots. He said that his interest is in constantly trying to make his robot do more things, and do them more intelligently. That brings to mind the giant robot of Benjamin Skora which is able (so they say) to vacuum the carpet, walk the dog, serve drinks to friends, and answer the door!

There will be more and more hobby robot kits available as time passes. They will be able to do more and more things, and will give their constructors hours of fun and pleasure putting them together, and watching them work. Like the radio controlled model airplane activity, the stage of hobby robot development is rapidly passing from the "built it all, figure it out, put it together, make it work" stage into the level where anyone can simply buy a kit and have it do almost anything the human mind can conceive of it doing.

REMOTE CONTROL OF HOBBY-TYPE ROBOTS

We refer you to the sketch of Fig. 13-1. It will receive feedback signals of voice, TV, mechanical movements, heat, light, and so on from the robot via radio communication. The robot will get its instructions and voice generation commands from the home computer by a return radio link. It is not such a long way off. The technology for this kind of operation is available. Modern radio control transmitters already are able to handle streams of pulses with reliability, and they are small and compact and do not use much power. They could easily be incorporated into a robotic form, or used at a home base station. The output of the computing section would modulate the transmitter, and in the robot we would find the second half of our communications system which would take the commands, use the memorized software programs, and cause the robot to do whatever we have planned for it to do.

MAGNETIC CONTROL

To make such a robot function around the home, it might be possible to use magnetic induction loops in rooms, and the robot could have sensors which would be able to pick up the signals from the centrally located computer in the house and do the things the computer directs it to do. Essentially we are saying that hobby robots, as well as commercial, industrial robots do require computer direction as they do things better.

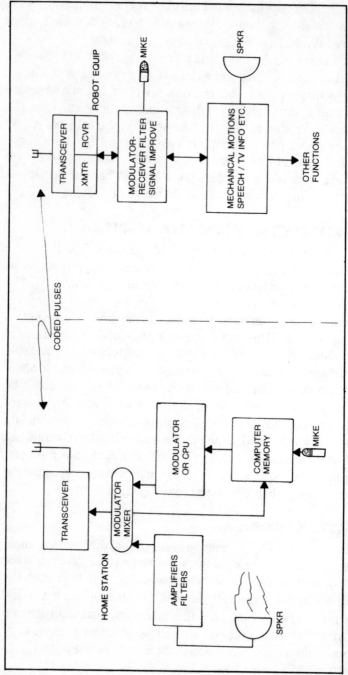

Fig. 13-1. A telemetry system for a robot's romote control.

THE MOUNTAIN HARDWARE SUPERTALKER: SD200

A talking circuit which was designed to be used with the Apple computing system is illustrated in Fig. 13-2. This system memorizes certain phrases and words and can use the memorized information to respond vocally to persons, under various conditions. This unit works with the Apple II Computer. It digitizes human speech and stores it, then it can play back this digitized speech through a speaker. The voice system consists of a card which plugs into a peripheral slot on the Apple II, and the mike, circuit board, and speaker shown. There is an extensive software package which permits the user to interactively develop a phrase diskette which may contain many tables of phrases. Each table can contain several words, phrases or complete sentences, which the computer can select.

Here is how the system operates. The Supertalker board Fig. 13-2, has electronic circuits which convert the analog

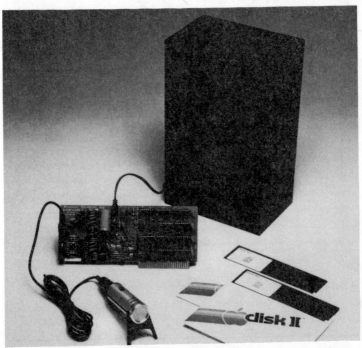

Fig. 13-2. The Mountain Hardware Supertalker system for the Apple computer (courtesy Mountain Hardware).

mike signals into a digital pattern which can be saved in memory units. The board's circuits accept these digital patterns and transform them back into analog signals for use by a loud speaker.

The circuit board has an on-board ROM that is a 256 byte type. This ROM is accessible from BASIC by a call instruction to address CN00 (Hex) where N equals the slot number that the board is in. Calling this address will place you in the talk mode, and a call to CN03 (Hex) is for the listen mode.

```
3   REM
5   REM      MOUNTAIN HARDWARE'S
7   REM  SUPERTALKER DEMO PROGRAM
9   REM
10  HIMEM: 4095
20  GOSUB 25000: REM  CALL AUTO SLOT FINDER
30  PP(2) = 16: REM  SET STARTING PAGE OF PHRASE TABLE
40  PP(3) = 127: REM  SET ENDING PAGE OF PHRASE TABLE
41  PP(4) = 3: REM  SET VOLUME (1-4)
42  PP(5) = 4: REM   SET SAMPLE RAT        E (1-4)
50  GOSUB 25010: REM  INITIALIZE SUPERTALKER WITH PP ARRAY
100 CALL - 936: PRINT
110 PRINT "HIT RETURN KEY WHEN YOU ARE"
120 PRINT "READY TO BEGIN RECORDING."
130 PRINT
140 PRINT "HIT RETURN KEY A SECOND TIME"
150 PRINT "WHEN YOU ARE DONE RECORDING."
160 INPUT A$: REM  WAIT FOR FIRST RETURN KEY
170 P = 1: GOSUB 25030: REM  CALL LISTEN SUBROUTINE
180 P = 1: GOSUB 25020: REM  CALL TALK SUBROUTINE
190 GOTO 100
25000 REM  TALKER AUTO SLOT FINDER AND ADDR CALCULATOR
25001 PP(1) = 1:P1 = - 16000
25002 IF  PEEK (P1 - 1) = 0 AND  PEEK (P1 + 3) = 3 THEN  GOTO 25004
25003 PP(1) = PP(1) + 1:P1 = P1 + 256: IF PP(1) < 8 THEN  GOTO 25002
25004 IF PP(1) = 8 THEN  PRINT "SUPER TALKER IS NOT IN THE APPLE II": IF PP(1) = 8 THEN
      GOTO 25009
25005 P3 = 256:P4 = PP(2): IF PP(2) < 128 THEN  GOTO 25007
25006 P3 = - 256:P4 = 256 - PP(2)
25007 P5 = P3 * P4:P2 = - 16256 + 16 * PP(1)
25009 RETURN
25010 REM  TABLE INITIALIZATION (ASSUME PP ARRAY IS SET UP)
25011 P3 = 256:P4 = PP(2): IF PP(2) < 128 THEN  GOTO 25013
25012 P3 = - 256:P4 = 256 - PP(2)
25013 P5 = P3 * P4: POKE P5,PP(4): POKE P5 + 1,PP(5)
25014 POKE P5 + 3,PP(3): POKE P5 + 4,134: POKE P5 + 5,PP(2)
25015 RETURN
25020 REM  TALK ROUTINE
25021 POKE 1528,PP(1): POKE 1656,PP(1) * 16
25024 POKE 1144 + PP(1),P - 1: POKE 1272 + PP(1),PP(2): POKE 1400 + PP(1), PEEK (P5
      + 3) + 1
25025 POKE 1784,(4 -  PEEK (P5 + 1)) * 16 + (4 -  PEEK (P5))
25027 CALL - 16384 + PP(1) * 256
25028 RETURN
25030 REM  LISTEN ROUTINE
25031 POKE 1528,PP(1): POKE 1656,PP(1) * 16
25034 POKE 1144 + PP(1),P - 1: POKE 1272 + PP(1),PP(2): POKE 1400 + PP(1), PEEK (P5
      + 3) + 1: POKE P2 + 3,(4 -  PEEK (P5 + 1)) * 16
25036 CALL - 16384 + PP(1) * 256 + 3
25037 RETURN
```

Fig. 13-3. One Supertalker demonstration program (courtesy Mountain Hardware).

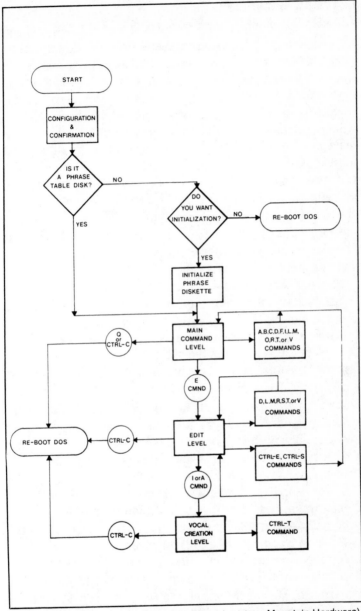

Fig. 13-4. The Supertalker computer flowchart (courtesy Mountain Hardware).

One Supertalker sample program is shown in Fig. 13-3, and the computer flowchart is illustrated in Fig. 13-4.

CHECKING ERRORS IN DATA
TRANSMISSION AND SOME ROBOTIC-SYSTEM PROBLEMS

Among the professionals, there are several methods by which data information can be checked and the capability of a system to transmit the bits accurately and completely is determined. One of these methods is to use a parity bit for each word or character and inspect this. Parity is a system in which we add a pulse to a coded group in such a way that for even parity, the sum of all the 1's will be an even number, and for odd parity, the sum of all 1's will be an odd number. In the transmission of an even parity code, those bytes which have an even number of 1's are not changed. In effect, a zero is added to the extreme left of the byte binary sequence. But, if the number of 1's in the byte is odd, then a 1 is added to the left to make the numerical total of 1's even. A byte with the wrong number of 1's has an error in it.

Other writers, in discussing the checking of data in transmission systems have indicated that a cyclic redundancy check might be used. This means, essentially, that you re-transmit the same information several times and if it all comes out the same, then the system has no errors—at least for that transmission period. Checks using special messages, such as the "Quick Brown Fox" which require transmission of characters representing all keyboard letters and numbers might be used. If, on the receiving end, one knows what the sequence of transmission is for this kind of special message, then errors in the transmission system can easily be noted. Finding the cause may prove difficult.

The number of methods of checking for errors in a computing system and its transmission system increase in proportion to the amount of information transmitted, and the number of people or stations involved in the generation, reception, and use of this kind of data. If one knows the particular weaknesses of his own system, which he learns by experience, then perhaps he can imagine and fabricate a checking system which will let him know how well his own system is operating. You need such a check in robotics, be it hobby type or industrial type robot system you are using, if it uses a computer and/or a communications system.

Checking a robotics system which involves mechanical movements, and computer control, and data and sensor transmission systems may be done using a test program. This program feeds data into the computing section to cause certain specified movements of the robot's arm in accord with the range of operations of which the robot is capable. If the robot is a welder or a paint sprayer, or such, then, of course, the test program must include some actual welding or painting or whatever to insure that the unit operates correctly from signal input to desired output. But, sometimes, as with computers, if there is a malfunctioning in the system, it won't respond as it should, and so location of the problem area is necessary although it may be difficult.

Piece-by-piece checking of a robotics system may be one answer to the error checking procedure. In professional circles this is called open loop checking. It is accomplished in this manner: First the end element is checked by feeding a signal into its nearest control point. One desires to eliminate all possible elements in between the signal insertion point and the end device and only those elements required are to be operated. Meantime all other elements of the system are turned off or disconnected.

When it is determined that the end element responds correctly to all the types of signals which experience or design or both show it should respond to, then one says that this part of the robot is working as it should. The next step is to move one element back toward the front-end or input of the system. When it is checked, of course the final element is also checked because we use the output of the robot as an indication of how well it is responding to test-input signals into the next-to-last unit. If there is an error or the robot does not do what it is supposed to do, and we know that the output element is working correctly because we have just checked it out, then we can say that the next-to-the-last unit is malfunctioning. We then check it out stage-by-stage or step-by-step until we find out what is wrong.

And so we move slowly back toward the input end of the system. At the input end we might have to use a conveyor belt to test input sensors, or devise other tests which simulate actual operational inputs of the same type that the input actually gets. We must test the input sensor system or input signal system just as completely as we have checked all intervening elements or system blocks. Finally, we might run a systems check. This is a closed loop test which actually is the robotic system in operation, doing what it is supposed to do.

In a robotics system, then, you'll check the computer section for accuracy and correct operation, you'll check the software to see that it was correctly prepared and inserted, you'll check the transmission system (cables, lines, and whatever) to see that the signals actually get to the operating units. You'll check the control amplifiers, and the motors and the hydraulics and the pneumatics and whatever. All this in a step-by-step procedure to insure that you actually will find the problem area.

If a system develops more than one malfunction simultaneously, then, unless you use the open loop test concept, you might not ever find the real causes of trouble! It has also been said by those who know about these things that sometimes two malfunctions make a no malfunctioning condition exist. The two malfunctions tend to cancel one another. This may happen if one depends too much on just a closed loop type of test.

Testing of hobby robots is performed much in the same manner. Usually you start at one end of the system, say the mobility end, the drive motors and gearing and power and such, and move back toward the input end of the system, checking as you go, to make sure everything is operating correctly. Lots of times this prevents troubles and helps you find and correct trouble spots which might not have been noticed, unless you check for correct operation as you proceeded backward, toward the input end of the system. It sometimes is very difficult to find trouble spots in any kind of system like this if you wait till you have it all built, and put-together, and then try to make it operate as it should. If it doesn't work correctly when completely assembled then you suddenly find there are ten thousand places where things might not be exactly right. Every solder joint, every cable connection, every switch, every mechanical device, and every electronic device suddenly becomes suspect.

The replacement of computer cards and circuit boards is a technique which, like current TV repair procedures, is standard. So, if you have a computerized robotic system, then you will want to make certain that its computing section can be broken down into small units which can be replaced if need be when a malfunction occurs and keeps on occurring and cannot be attributed to such things as transients, lightning, or static.

Through the courtesy of Ohio Scientific Company we show, in Fig. 13-5 an example of their computer's circuit board with its attachment cabling, and even, to the left, some internal battery powering units. Notice that neatness and relatively good spacing between integrated circuit chips are predominant in this fine engineering example.

THE TRS-80 AND THE OUTSIDE WORLD

We examine the use of the TRS-80 in a control application, courtesy Radio Shack and we can, from the information gleaned, expand our imagination to extend to other applications of such a computer in a robot. It is indicated, in instructions on the use of the TRS-80 in this control type application that external circuitry is required, and also that you will have to have proper software to make the control functioning possible. If you have a Level I machine then you'll have to write the software program. If you have a Level II machine, then you can use the POKE and PEEK or the OUT or IN machine instructions.

It is said that when you design the hardware (circuitry) for computer control of something, you need to have thought out your software program completely so that the two are compatible. How you design your hardware will be determined by the instructions you use to operate that hardware. There are two operational approaches with the TRS-80 (and perhaps other similar microcomputers (which can be used,

Fig. 13-5. The Ohio Scientific's computer circuit board (courtesy Ohio Scientific).

these are the memory mapped system (Fig. 13-6) and the second is the port system (Fig. 13-7).

If you use the memory map technique then you must specify a memory address that will be the location of the hardware circuitry controlling your device(s). To write data to your control system (send it instructions) you address it via the address lines and send it the instructions over the data lines.

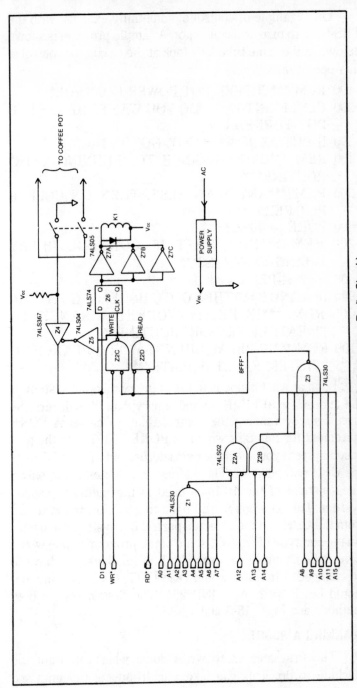

Fig. 13-6. Memory mapped coffee pot control using the TRS-80 (courtesy Radio Shack).

435

One example of control of something external with the TRS-80 is to turn on a coffee pot. A sample program is shown below. At the same time let's look at the circuit for control of this operation.

```
100  REM ***COFFEE POT POWER CONTROL***
200  CLS : PRINT@512, "DO YOU WANT THE COFFEE
     POT TURNED ON";
300  INPUT A$: IF A$ = "NO" GOTO 7840
400  REM ***IF NO, BRANCH TO "RUNNING LATE"
     ROUTINE***
500  REM ***ANYTHING ELSE, TURN ON COFFEE
     POT RELAY***
600  POKE — 4092,2
700  REM ***NOW TEST IF CONTROL RELAY
     CHANGED STATES***
800  B=2: A-PEEK — 4096
900  IF A AND B = 0 THEN GOTO 1980 ELSE GOTO 3744
1000 REM ***IF RELAY WORKED, BRANCH TO
     "WEATHER SENSOR" ROUTINE**
1100 REM ***IF RELAY DID NOT WORK, BRANCH TO
     "SYSTEM FAULT ISOLATE"***
```

Look at what happens in the memory mapped system. If the address is 8FFF (Hex) and binary data 02 will turn the circuit *on* by causing the control relay to close. A POKE statement for this task would be POKE 36863,2. In the port based system you specify a port address out of the 256 ports the CPU will address. The address is selected using only 8 lines instead of the 16 lines used in the memory mapped system. The data bus is still used to get information to and from the CPU and the selected port. If the system had used a port based type of control, some of the program above would have to be changed, for example: Lines 600,800 will have to be changed. Line 600 would read OUT 254,2 and line 800 would be: B = 2: A = INP 254. The flowcharts for both methods are Figs. 13-8 and 13-9.

PLANNING A ROBOT

The first step is to write down what you want the machine to do. This can take a multitude of thoughts and

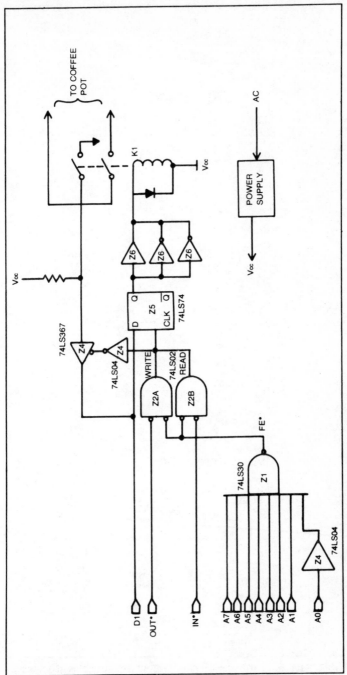

Fig. 13-7. Torque vs wheel size considerations for robots.

437

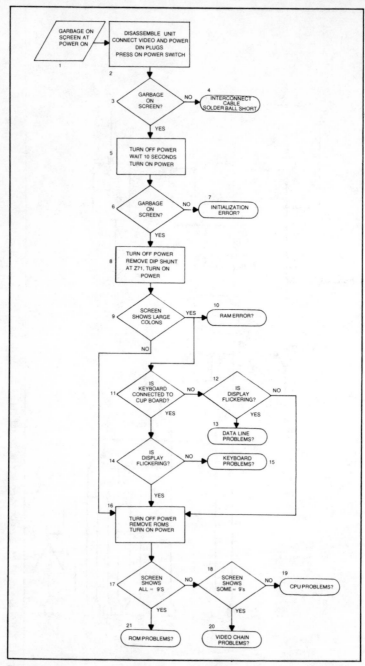

Fig. 13-8. Section isolation flowchart for TRS-80 (courtesy of Radio Shack).

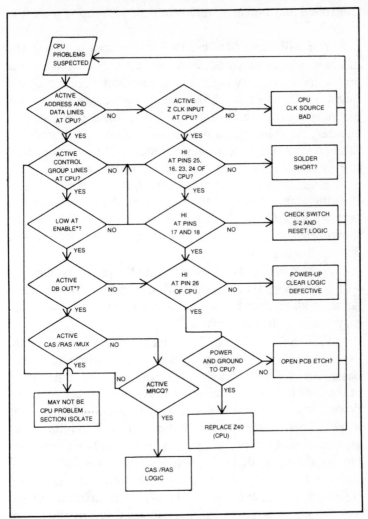

Fig. 13-9. CPU flowchart for TRS-80 (courtesy of Radio Shack).

statements in some cases, and only one statement in other cases. For example, one person had the initial objective "I want a robotic machine which will entertain people." And from that basis he did construct a robot which did exactly that, and delightfully so. There is quite a difference in what you will want to incorporate into the machine if it is for entertainment rather than being built to do some specific task. One machine might be much more complex than another, and how it is controlled may be vastly different depending upon what it is to do.

You will immediately add that the appearance of the machine will, or might be, governed by what it is to do, and you would be exactly right. For entertainment purposes we might want a robot which looks manlike, or girl-like, but for job purposes, such as taking objects from one place to another, the robot might better resemble a small car with a flat bed on it and have somewhat different arms which load and unload the material, than those found on the kind-of-human-duplicate the entertainment robot might be. So, when thinking about what the machine is to do, or to state the purpose for building the machine, you will probably also be thinking about what it might look like in the finished version. This is a second step.

Next, and closely associated with the previous thinking, will come those thoughts related to mobility. We can make some doodling sketches as we have done in Fig. 13-10 which may assist us in our thinking.

You can expand on this concept and come up with, perhaps, a dozen different possible illustrations of how a baseplate of a robot might be made and how it might be propelled. In our illustration at A we show a conventional drive using a common axle and one drive motor for one rear wheel. This is important because you can't get the robot around a corner if you make both rear wheels a drive type. One must be free, that is, its hole slightly larger than the axle so it can turn freely and on its own. The other wheel is fixed rigidly to the axle or so mounted that a gear can be attached to it and this gear is meshed with the motor drive gear to propel the baseplate (and robot). The steering is just like any automobile steering, or toy car or truck, using a turning motion for both front wheels which are linked together. This will work, but the robot may have some problems because the turning radius of the base is large. It is a stable base, however, and can carry lots of weight as we know.

At B is a diagram showing two independent motors driving two wheels, and use of small idler wheels, front and rear, to give the platform balance. This type system is com-

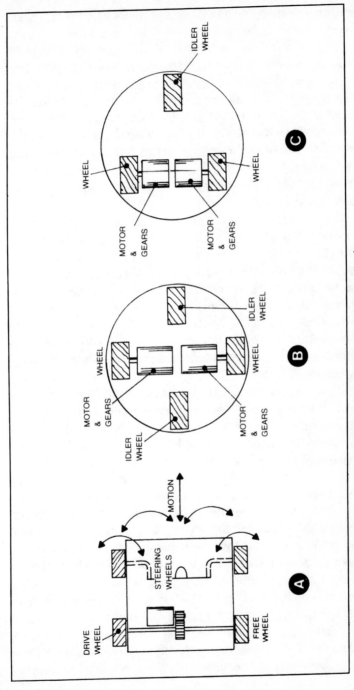

Fig. 13-10. Some ideas and sketches of possible robotic base-structure mobility methods.

monly used when you have stepping motors for drive power. If you step the drive motors at the same speed the base goes straight forward (or backward). If you step the motors at different speeds, a turning motion results. Stepping one motor forward and the other backward at the same speed gives a very tight turning circle. It is easy to synchronize the speed of stepping motors just by observation. If the robot starts to turn somewhat as it goes forward, you just move the control (assuming radio control) of the slow motor to increase its pulsing rate slightly and synchronism is quickly obtained. The GARCAN robot which we discussed in an earlier chapter uses this system.

At C of Fig. 13-10 you see how one idler wheel has been removed. The platform is not as stable as before, but can be all right for level floor operation. Also, if you want to try to use the idler wheel as a drive wheel, this is possible, and you can so arrange the mechanics of the system that the one driven wheel can be turned for steering purposes. You might use a single motor in that case. Or with the three wheeled configuration you might use a single motor to drive both rear wheels as in (A), or use two stepping motors. There are many variations.

So, you doodle and try to come up with some kind of a base-and-drive-and-steering system for your robot, if it is to have mobility. Of course, not all robots (domesticated types) will need mobility. They may be permanent fixtures in the house wall section somewhere, yet they can control all kinds of functions around, and in, and from the home. These are decisions which you must make in your planning.

You can select a wheel size and from that as a starting point calculate the gearing needed and the motor horsepower and so on. You select the wheel diameter based on where the robot is to move. If the surface is smooth and has little change in level, then a relatively small wheel size, say 4 to 6 inches diameter might be all right. But, if you plan the robot going into grass or on rough terrain or unpaved streets and sidewalks and such, then a larger wheel diameter is necessary, say 12 to 18 inches might not be unreasonable. Some may argue this point so we won't say it is required, we just say the larger the wheel diameter the easier it is for the wheel to get over small rocks, and thick grass, and earth depressions and indentations.

But there is a torque penalty as you increase the size of the wheel which is driven. In Fig. 13-11 we show this relationship, again on a doodling basis. In A you see a large wheel with a given size gear drive. Notice that the forces retarding motion will be the resistance or opposing force (torque) times the lever arm to the wheel rim from its axle. Now the motor torque must, at least, equal this value, and here its lever arm is much smaller, so where do we get the higher number to make equality? We have to get a larger motor torque value than the opposing torque value. The length of the torque arrows (vectors) illustrate this. Now examine B where a smaller wheel is shown, the lever arm is smaller for the opposing torque so the frictional retarding force may be less, thus the motor drive torque needed is less. This is one effect of using a large wheel vs. the use of a small wheel. Notice also the rock in the path of the wheels. For the small wheel it is a high obstacle, for the large wheel, it is not too large an obstacle. The larger wheel may go over the rock, the smaller wheel may not.

In our planning, then, we come up with something as to its size, what it is to do, how we will steer it or control it, or have some notion that it might steer itself, and then we doodle a little, trying to see how and what we want to put together to make it, and from this initial effort we find out other things we need to research and look into to determine how we will put our robot together. We will have looked at where it will operate.

AFTER MOBILITY AND STEERING, THEN WHAT?

That's a good question. You might not have finished with the steering planning, even if you have decided, approximately how you are going to make your robot move and what power supply you'll use and so on. You might be thinking that you'd like to have the robot autonomous, and so that means that some of the steering, at least, must be done by a computer type electronics device. We don't say a computer di-

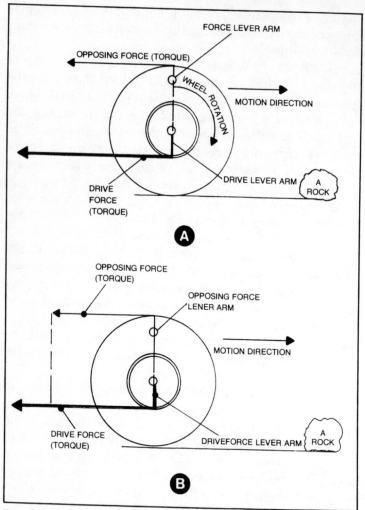

Fig. 13-11. Torque versus wheel size considerations for robots.

rectly because most have display screens, and maybe disk memories and such, which you actually may not need. They also have keyboard inputs which may or may not be appropriate to your planning. Actually, what you may want for your robot may be more of a comparison type system with memory banks, which can be programmed, and with sensor inputs from all the various devices you imagine might be useful to the machine. It would be nice if you have a kind of plug-in keyboard which you can connect to the robot via a cable, and through which you can adjust its memory. By adjust we mean program, erase and re-program, as you make command changes, and put in directions.

We have examined how a robot is taught in an industrial application. He is "programmed" through certain maneuvers using a small hand-held controller. The electronic circuits are so arranged that as long as the arm is in movement, the computer does not remember what it is doing, but when the arm stops for any period of time, the computer makes a note of this. Later, when the arm is under autonomous computer control, the computer sends the arm to the various end points precisely, quickly, and directly and the arm path may not be the one *you* used to get it there. It might be possible to easily program a hobby robot much in the same manner but using a radio-control system as the hand-held controller. Suppose we consider that in your robot you will have a radio control receiver whose output goes to a computer and various devices to be controlled, Radio-control systems such as are used in model airplanes can be obtained with up to 8 or so channels. If you need more than that, use two systems, operating on different radio frequencies.

Of course you might also use a hand-held controller with lots of push buttons, attached to a cable system which can plug-into the robot's body, and you can follow the robot as you command it to do whatever you want it to do. The signals could be recorded on some kind of tape, or placed in the memory of a computing section or whatever and then played back later.

MORE CHEAP SPEECH

To know what is on the shelf in the way of equipment which you might use in a hobby or industrial type robot is often very helpful. Thus we call attention to the Radio Shack Voxbox and Voice Synthesizer shown in Figs. 13-12, 13-13. The Synthesizer allows the computer to "say" anything built from phonemes. The Voxbox is designed to work with the TRS-80 computer and is designed with a 32 word memory which permits it to remember 32 words which you teach it,

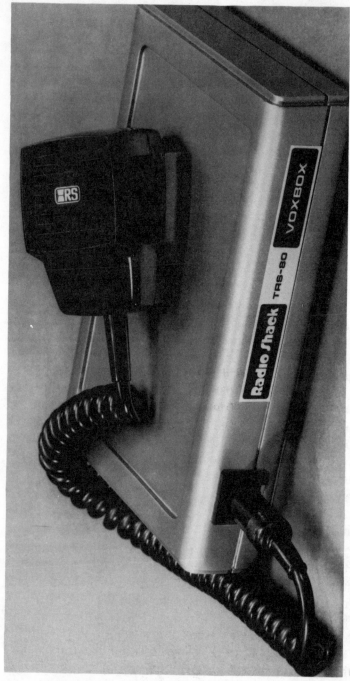

Fig. 13-12. Radio Shack's Voxbox system (courtesy of Radio Shack).

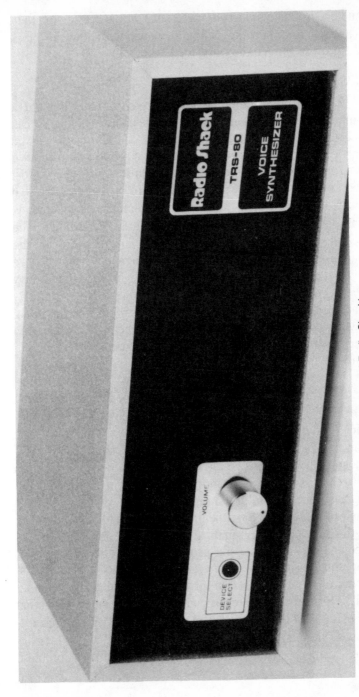

Fig. 13-13. Radio Shack's Voice Synthesizer system (courtesy of Radio Shack).

447

and produce a logic output when such words are spoken to it. Through the courtesy of Radio Shack we show the block diagram of the Voxbox in Fig. 13-14.

COSTS AND ACCOMPLISHMENTS

It has been said by many who are experts in the field of human studies that humanity cannot be happy unless they can accomplish something. In fact, this is one fear of those who think about such things—that the age of robotics will deprive persons of doing something worthwhile—in other words, working. Every job performed, in spite of its problems and frustrations and exhausting demands, means that someone has done something which they consider worthwhile. The sociologists think that this, perhaps, has meaning for the happiness and contentment of the human race. Thus, when we evaluate what we have when we have finished our hobby robot, whatever type it may be, we must evaluate our efforts in more terms than just of money or time spent. We must also, always, evaluate in terms of what has been learned and in terms of pleasure received, and personal satisfaction obtained.

There are very few if any, persons, who will say, directly, that the future does not hold a place for industrial robots. There are too many foreign countries developing such devices for use in their areas of productivity. And with such success that our own industrial leaders say that a clear view of those operations. Much planning and consideration must go into the concept of using more automation in existing plants, because of the worker commitment to the various jobs. Yet the profit motive, being the primary reason for the existence of the plant or industry in the first place, will demand the most economical, efficient, and productive operation possible at the smallest possible costs. Let us examine what might be some of the consideration for the use of industrial robots.

Do they contribute to productivity, efficiency, reduction of error, a better product, faster production, reduce hazardous situations, reduce tedium, improve the work environment, reduce costs? In many cases this requires a complete

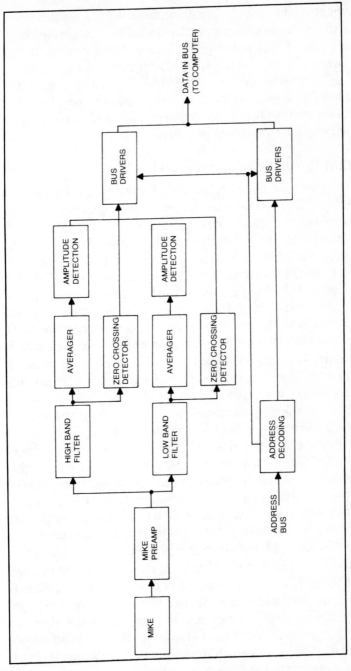

Fig. 13-14. The Voxbox block diagram (courtesy of Radio Shack).

review of the whole operation, down to the most minute detail, of the business, or plant, or industry wherein the use of industrial robots is being considered. We are in mind of a system of plan evaluation or job operation planning called PERT (Program Review and Analysis Techniques) or, as it is called by some, the critical path planning approach to—whatever job or task or operation is in mind. Can it help show the need for automation or robotics?

WHAT IS THE CRITICAL PATH APPROACH

If we take the start and end of a series of tasks and call them events, and we take the activities in between the start and end of each such task and call that an activity, we can make a diagram showing this concept as illustrated in Fig. 13-15. Notice that there are many activities (arrows) shown, 5 in total. Also know that it will take the completion of all 5 activities before the end event can be said to be done! At this point we do not know from whence the other activities originate, or how they may be inter-related, or interdependent upon other end events, but be assured that in our modern industrial world, there will be a rather complex network before we come to that stage where the end event actually means the completion of the job!

The starting event can be merely a time and date at which some activity begins. The end event can be a time and date when the job is finished. The activity can have a length which is somehow related to the time it will take to do that activity, based on the educated guesses of many experienced personnel who have some knowledge and, hopefully, experience with doing that kind of activity. Thus we might say activity #3 will take 4 days to complete, by best estimates.

Think what this means if it is required that all the other activities must be done within that same time frame in order for the end event to properly occur on time! As you might look over the time and date estimates of each activity, you might find one which is so long that it is a bottleneck. It will be that long-time activity which will hold up the job which needs some attention. The path through that activity from its starting event, to the end event, is the path known as the critical

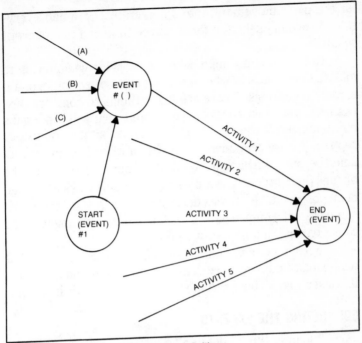

Fig. 13-15. Some pert or critical path definitions.

path. It is a trouble area for some reason. It is stated that this critical path is the one which requires the most time for any process or object to get from the initial event to the end event. Also, it is stated that if there is any further delay along this critical path, then the final event will be delayed by whatever time the delay on the critical path assumes.

So, how do we relate this to robotics in an industrial sense? If we consider an example where lots of parts must arrive at a given station at a given time in order for assembly to be accomplished, and that a delay is found on one path, we will assume a simple case here, wherein the distance the part has to move, and the processes which have to be accomplished on this part, are responsible for that delay, then we can see how an automated situation may remedy the problem. It is not unreasonable to think of robots doing the processing, and perhaps saving time doing so. It is not unreasonable to think of robotic conveyors transferring the parts more quickly. Thus we change what was a critical path for the parts movement, into a non-critical path and everything is accomplished in the required time and lots of profit results!

There are many approaches to consider when trying to make a decision as to whether or not to robotize some, or all, of your operations. There are many things to consider. For example, you must always have in mind what will happen if a machine fails. No one likes to think of this, but it can and does happen. We are reminded of a situation wherein some robots failed because the human attendant did not keep the hydraulic oil tank filled properly, nor did he (she) report leaks which had been observed, etc. It was determined that it was that worrisome case of a human hating a robot and thus hoping it would cause trouble and helped it to do so!

There is no question but that, in some cases, the robot will displace humans in certain work functions and activities. It is just an axiom but for some things, robots can do it better.

CONSULTING THE EXPERTS

Of course this is done. One just doesn't expect all the people in his own operation to, necessarily, know all about robots and robotic installations, and operation and maintenance and what can and what cannot be done with these machines. Sometimes, and this is true in most cases, an analysis of your operation by a robotic expert will help to show what should or should not be done. However, one must always consider that probably any representative from any robot manufacturing company will be a salesman. But most will give a good, honest, and reliable report on the situation at hand.

Because a conversion to industrial robots is an expensive operation, probably in more ways than just changing equipment, management will also want to inspect and discuss robot plants with those who have already done this in part, or, as some Japanese have done, completely robotized a plant or industrial operation. One question which always comes to mind is, "What do you do if some machine fails or stops, and it is a line item which must function in order for the whole operation to function?" First, of course, there needs to be an alarm system. We have mentioned this in some previous pages while discussing industrial robots. The alarm can be of several levels. One, a warning that trouble is imminent, and where it might occur, second, that trouble is happening and where and why (sensors can determine this in many cases), and what should be done about it: a rerouting, stopping all machines, or whatever is appropriate, even to the extent of human intervention and human substitution for the machine until the robot is fixed.

There are many other items which need be considered, such as a loss of electric power, even for just a short period of time, and how this might affect computer programs, movement, and synchronization and so on. One might use the operating room technique wherein a substitute power plant comes on immediately in case the main source of power starts to fail. When consulting the experts in the field, one will become knowledgeable of these various situations and problem solutions for these situations. It is also considered good management if one can obtain the help and advice of the normal plant complement of persons, who can give suggestions which will make their job work easier or more palatable, and thus they provide insight into possible areas of automation which will be accepted by them. Education and instruction of plant personnel is, of course, mandatory. The experts can help with this type of public relations.

ROBOTS AROUND THE WORLD

It is also interesting to consider the robotic development of some foreign industrialists. A study by Professor Gustav Olling of Bradley University states that Japan has 13,000 of the world's 17,500 industrial robots. They have some 70 companies who are developing new robots and this compares with some 27 companies in the U.S. It has been reaffirmed that the Japanese goal is still the unmanned, robot operated factory.

The progress in industrial robotics has been remarkable and will continue at an astonishing pace. It has been said that the new robots are simply programmable arms with some high degree of manipulative skill. But that they have poor senses and a somewhat low intelligence which is not yet able to comprehend and really handle the information which is presented by the various types of sensors. But that even so they are most valuable in industry. It is not uncommon for industrialists to recover their capital investment in robots in somewhat under three years.

It has been pointed out that robots need something better than the type of grippers which are now common. But these improvements are under research and development, especially in Japan and in Russia. Robots which have fingers that can handle objects as delicate as an egg without any danger of rupture, and quickly adapt to different and varied object forms have been made. It has been stated in various studies that one of the most valuable ways a robot can improve plant production is by speeding materials through a plant. It was found that materials are worked on only some 5% of the time, and the rest of the time that material is lying idle or being transferred from one place to another. When robots are specifically designed for a task, they can offer some very remarkable work. For example, the Fiat Robogate, installed in its Rivolta and Cassino factories is said to be able to accomplish all the required welds on a car body in less than one minute. There are more than 50 robots on each welding bay. The precision of each job is the same so that quality control problems are reduced, and should improvements be needed, a slight change to the control program affects the desired result.

The use of robots on an assembly line seems to be inevitable and no matter what the task, it now seems possible that the job can be done by such automated machines. The human system seems to break down when subjected to the constant boredom of doing the same task over, hour after hour, day after day. It was found that in Silicon Valley of California where chips or integrated circuits are developed and made, that this type boredom has led to such advanced human problems as drug usage. That, of course means profit loss to management in that quality is sacrificed, and integrity of the firm or plant is jeopardized. That means a loss of buyer confidence. Again, here, in this type environment and task

454

situation the robot type machine offers no such problem. It does what it is programmed to do (what it has been taught) endlessly, hours on end, without error, loss of quality, or complaint.

As robotics components are developed and improved, such as the new TV cameras-on-a-chip developed by Hughes Aircraft Company using the new advanced charge-coupled technology which we have discussed somewhat in earlier pages on robotic eyes, the ability of the robot to do more things better will be almost beyond imagination. These Hughes Omneyes, as they are called, consist of a chip of some 1,024 light-sensing elements for one type application, and another chip with some 10,000 picture elements for higher resolution. It is envisioned that the ultimate use of these robotic eyes will be on the assembly lines to help the machines size, orient and identify parts and objects. Because of their small size and low power requirements, they can be used close-up if necessary on smaller and smaller type objects, perhaps ultimately, even down to integrated circuits themselves.

JOBS

When one considers the use of robots in a factory or plant the questions and job-loss fear of employees, and Management's commitment to the workers, becomes very real situations to deal with. Some questions asked by workers are, "Will robots in factories eliminate jobs?" The answer, of course is, yes. The robots will eliminate a majority of the old, dull, hazardous, and non-progressive jobs. But, if one is willing to learn and adapt to a new world of robotics, then there are opportunities which one should consider. The following comes from a study conducted on Careers in Robotized Factories.

Robots need repair sooner or later. Robot repairmen will be in high demand and will be very well paid. These personnel will also check and test the robots to insure maximum performance with a minimum of breakdown. There will be robot trainers, i.e., those who program the robots, and that will be a field in which computer knowledge and new skills in machine programming will be learned at special schools provided by the user plant or the robot manufacturer. There will be those who will be responsible for moving the robots to desired workplaces if the robots are the fixed location types. These persons will coordinate the robot's work load, and will see to it that they are productive at their new line locations. Of course there will be more and more need for robot designers. These persons who have engineering backgrounds in control and electronics and mechanical engineering and physics will obtain very high salaries and find their work challenging and very productive. Industrial robot use and plant analysis may be another rewarding career field in which people, sociological problems and machinery are all considered together. It is going to be a long, long time before robots will be able to repair and adjust and teach themselves much more than very elementary operations. So they will need lots of care, adjustment and help.

The respected *Wall Street Journal* has reported that some factories in which robots are built by other robotic machines without human control, have been under construction by some Japanese electronics firms. This has been the beginning of a deep seated fear of job loss among many people who are not knowledgeable about such machines. It is entirely feasible to specify, down to the smallest detail, the parts and layouts and connections of those parts, to make up a robot, and then to design machines which will do the assembly work so that at the end of the assembly line, a so called robot will appear, ready to go to work.

Are robots on the increase in industry? Ransberg and Renault have formed a venture to make and deliver industrial robots with each system selling at $100,000 to $200,000 each. It is said that they have invested some 15 million dollars into this venture. The chairman of the Ransburg Corp. (France) has estimated that the U.S. market for industrial robots is expected to be in the vicinity of $500 million dollars annually by 1985.

If there is a process in a factory or plant or industrial complex which cannot be robotized, it is just because no one has examined it yet for that purpose. The many fine robotics companies are leaping ahead, year after year, with advanced and improved designs as the technological explosion continues, and there seems to be no end in sight. It may well be that in the not too distance future, factories and plants and industrial complexes and shops and stores and assembly places and whatever will have to consider using robots in self defense—in self defense of profits, that is.

THE COMPUTERIZED HOME AND ITS SECURITY AND CONTROLS

The computerized home might also be called a robotic home. What is a robotic home? It is a home in which there are many automated devices which make that home more secure, easy to keep, and pleasurable for you to live in. In a sense this means that you will make your entire house a kind of robot and you will live in its protective embrace.

On the practical side, of course, what we mean is that your home will have some kind of centralized computer which can sense and control a multitude of functions in and around the house. One such computer is the Scientific Instrument's C8P which we use for an example here, although there are many other types which can also do this kind of job (a job is an action requiring many tasks) for you. Briefly, here are a few other control items.

With the speech capability, the person might not have to touch a bar at all. They simply say "Return" and this activates the return program in the computer. Or they could say "Help", and this would activate another computerized subprogram which would sound an alarm and indicate where the trouble exists. We have previously stated that such things are possible. We look at Fig. 13-16 to see one example of a voice system produced by Ohio Scientific in their CA-14A system. system.

When you want to make things happen in the house, such as to turn on the coffee pot, and turn on and off various lights and other electrically operated devices, you need a kind of special type computer unit and remote controlled modules. One example of this type system is shown in Fig. 13-17.

Fig. 13-16. Votrax, a voice system produced by Ohio Scientific in their CA-14A system.

The master control unit is so programmed that when you press down the various buttons you can cause the circuits controlled by the modules to energize, thus you turn on lights at will from a remote location. Again, using the button controls, you can deactivate those remote circuits also. Since the control signals are sent over the house wiring, it is no task at all to install such units, you simply plug-in the remote modules into a convenient electric outlet, and run the cord from

Fig. 13-17. AC-12P wireless remote-controlled system (courtesy Ohio Scientific).

459

the master box to the outlet in your remote command location. You plug-in the devices you wish to control into the modules receptacles and you are ready to operate. By the way, don't do as some do, they plug in lights but forget to turn the switch on. When the signal is sent from the remote location to energize that light (or whatever), it won't work, because its own switch is off. There is a unit made that interfaces this system to a TRS-80.

In this robotic home concept we must not forget the possibility of remote control, from some pleasant beach resort to our homes back in our own home town. Yes, it is possible and feasible when we use a device called a telephone interface such as is shown in Fig. 13-18.

Looks complex, doesn't it. It is. But that is no problem to us. When this is connected so the telephone has an input to it you can communicate with your home computer from almost any place in the world. If your telephone system is a Touchtone phone you can obtain status reports on various things in the home when you send certain coded groups of numbers which the computer can recognize and respond to. For example, you can tell the computer to water your plants, give you a report on the integrity of your windows and doors, test the air for smoke or other bad odors, also to play back for you, any recorded messages which are on file in the telephone message recording machine.

With appropriate sensors you might get a report on the grass height and the dryness of the ground. If the grass needs watering, then you can tell the computer to water, using the robotic sprinkling system, for a specified period of time. Imagine also that you found the grass too high, in your status report. You simply send in a new series of coded numbers and the computer will order the robot lawnmower into action. It will cut the grass, following a carefully prearranged routine path route and then return itself to its garage home or wherever. It could be that in the future all mail will be electronic. When you query from Hawaii or elsewhere the computer will tell you to pay such and such bills and ask you if it is all right for it to issue the checks! Yes, 'tis true, all things are possible nowdays in our computerized, robotic world.

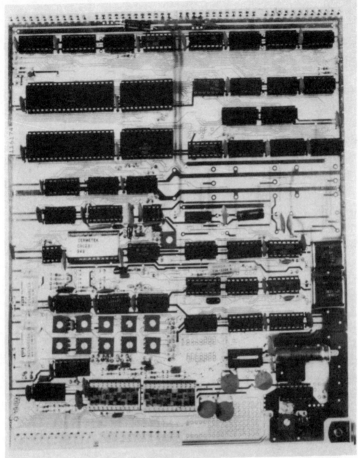

Fig. 13-18. CA-15 Universal Telephone Interface (courtesy Ohio Scientific).

SOME STEPS IN ROBOTIC DESIGN

Be it hobby type or industrial type, there are certain steps which are generally followed in the design of robots. These can be listed as follows:

(a) List the functions desired to be accomplished—what it is to do now and in the future.

(b) Draw block diagrams showing the interconnection of various units such as drive system, signal processor, etc. Use as many block diagrams as necessary to come up with some kind of complete picture of the machine.

(c) Check the availability of the parts needed for each block of your block diagram. If some parts are not available then improvise, invent, construct, or change the system in that block.

(d) Construct breadboards of your circuits and test them, adjust them and modify them until you get the necessary and correct operation from each board reliably and repeatably.

(e) Construct models (miniatures) of all moving, working parts. Check for construction difficulties, needed materials, and make such changes as necessary to accomplish the end objective for each unit. Test and adjust each unit.

(f) Construct a full size working model of the robot—leaving room physically for changes and modifications—being sure you have easy access to parts and test ports. Check the operation of each unit and the complete assembly.

(g) Construct the final package which has the desired shape, size, and appearance you want. Test and adjust it. Operate it, or let it operate itself. Check for faults and reliability. Study it for improvements. Then, have fun!

Index